Springer Praxis Books

Astronautical Engineering

More information about this series at http://www.springer.com/series/5495

Donald Rapp

Use of Extraterrestrial Resources for Human Space Missions to Moon or Mars

Second Edition

 Springer

Published in association with
Praxis Publishing
Chichester, UK

Donald Rapp
South Pasadena, CA
USA

Springer Praxis Books
ISSN 2365-9599   ISSN 2365-9602   (electronic)
Astronautical Engineering
ISBN 978-3-319-89196-5   ISBN 978-3-319-72694-6   (eBook)
https://doi.org/10.1007/978-3-319-72694-6

Printed on acid-free paper

This Springer imprint is published by Springer Nature
The registered company is Springer International Publishing AG
The registered company address is: Gewerbestrasse 11, 6330 Cham, Switzerland

Preface

Since the earliest expeditions of humans into space, visionaries have contemplated the possibility that extraterrestrial resources could be developed and civilization could eventually move into space. An important early paper (Ash et al. 1978) essentially opened up the realm of ISRU. They proposed that propellants for ascent be produced on Mars. Thus, the term *In Situ Propellant Production* (ISPP) was coined, and this provided a focus for a couple of decades. ISPP on Mars was the most obvious choice for the utilization of extraterrestrial resources because it provided very high leverage and it appeared to be more technically feasible than most other possibilities.

As time went by, visionaries looked beyond the near term and imagined the transfer of the industrial revolution and the electronic revolution to planetary bodies. Metals would be produced and fabricated into objects, concrete building blocks would be assembled into structures, and electronics would be created from indigenous materials. Eventually, ISPP became an obsolete term and it was replaced by *In Situ Resource Utilization* (ISRU) to allow for a wider range of applications than mere propellant production.

Robert Zubrin is a prominent Mars technologist and advocate of Mars exploration and is Founder and President of the Mars Society. His book *Entering Space* provides an impressive road map for humans to settle in space.

Zubrin contemplated finding "fossils of past life on its surface," as well as using "drilling rigs to reach underground water where Martian life may yet persist." He believed that there is great social value in the inspiration resulting from a Mars venture. He also said:

> The most important reason to go to Mars is the doorway it opens to the future. Uniquely among the extraterrestrial bodies of the inner solar system, Mars is endowed with all the resources needed to support not only life, but the development of a technological civilization... In establishing our first foothold on Mars, we will begin humanity's career as a multi-planet species.

Zubrin has support from a good many Mars enthusiasts. (The goal of the *Mars Society* is "to further the goal of the exploration and settlement of the Red Planet.") More than 10 years ago, it was claimed that we could send humans to Mars "in ten years" and begin long-term settlements. Each year, the *International Space Development Conference* hosts a number of futurists who lay out detailed plans for long-term settlements on Mars. The *Mars Society* often describes settlements on Mars as the next step in the history of "colonization" and warns not to make the same mistakes that were made in colonizing on Earth. For example, the Oregon Chapter of the *Mars Society* said:

> When the initial settlements are set up, there will most likely be a few clusters of small settlements. As time goes on, they should spread out. The more spread out the developing townships are, the more likely they will develop their own culture. In the beginning, townships will be dependant [sic] upon each other for shared resources, such as food, water, fuel, and air. Once a more stable infrastructure is set up on Mars, then people should be encouraged to set up more isolated townships. In any area were colonization or expansion has occurred, one important item that cannot be ignored is the law. Some form of law will be needed on Mars. Looking at the system that was used in the old west, we can see that whoever enforces the law can have difficulty completing his job. The 'sheriffs' on Mars must be trustworthy individuals that the majority of people agree on. They should not be selected by the current form of politically interested members of society; this only encourages corruption. Instead, some sort of lottery system of volunteers should be allowed. As for the law itself, it should be set in place to guarantee all of the basic rights of everyone, from speech to privacy.

While these zealots are already concerned with establishing law and order on Mars, and spend time laying out townships for the Mars surface, this humble writer is merely concerned with getting there and back safely and affordably.

ISRU visionaries know no bounds. Imaginative proposals abound for all sorts of futuristic systems. One example is a sort of Zamboni vehicle that rolls along the surface of the Moon or Mars, imbibing silica-rich regolith, processing it into silicon in real time, and leaving in its wake a roadway covered by a carpet of silicon solar cells that stretches out for miles behind the Zamboni.

NASA is not a monolithic organization. Imbedded within NASA is a small cadre of ISRU advocates who have urged NASA to support Mars ISRU technology because of the very significant mission benefits. They have sought support from the greater NASA for further development of ISRU, but such support has been intermittent and inconsistent. Since the 1990s, ISRU advocates have developed elaborate plans for the development of ISRU technology but NASA funding has been concentrated into a few relatively expensive demonstrations while technology funding has been limited and intermittent.

Higher NASA management has provided vacillating leadership over the years, with programs and initiatives sometimes starting with great fanfare and ending abruptly without warning.[1] Budgets rise and fall, and continuity from year to year is difficult to achieve. The greater theme of NASA technology has evolved from *unprecedented*, to *world shaking*, to *revolutionary*, to *game changing,* to *disruptive*.[2] The focus of

---

[1]This brings to mind the six stages in a NASA project: (1) wild enthusiasm, (2) great expectations, (3) massive disillusionment, (4) search for the guilty, (5) punish the innocent, and (6) promotion for the non-participants.

[2]As a result, the NASA technology programs have often been *disrupted* because the *game has changed* so frequently.

technology has always been on seeking incredible breakthroughs, and therefore, funding to do the engineering necessary to make evolutionary systems practical has not usually been forthcoming. The focus of human exploration has been to occasionally fund a Mars demonstration project. Visionaries tend to look beyond the best near-term prospects. In this environment, at each juncture when a new technology opportunity arises, the tendency is for NASA ISRU managers to ask NASA HQ for far more funding than can reasonably be expected, and hope to get some fraction of what was asked for. Inevitably, the long-range plan is so over-bloated with ambitions that the divergence between actuality and the plan becomes embarrassing. In 2005, when the NASA *Vision for Space Exploration* was announced, the ISRU enthusiasts wrote plans for extensive robotic and human precursors to validate ISRU on the Moon and Mars, none of which were ever funded, nor was there any serious reason to believe they would be funded. The entire exercise, like most planning activities for ISRU, was basically fiction and fantasy. When the whole NASA enterprise was diverted to lunar mission planning, the small amount of work attributable to Mars ISRU was canceled and new funds were allocated solely for lunar ISRU research.

Unfortunately, lunar ISRU in any form does not seem to make much economic sense to this writer. Furthermore, the technical challenges involved in implementing lunar ISRU are immense. None of the lunar ISRU schemes appear to have a practical financial advantage and it appears to be better, cheaper, and simpler to bring resources from Earth to the Moon—at least in the short run. By comparison, some forms of Mars ISRU have the potential for great logistic and financial benefit for human missions to Mars. Yet funding for Mars ISRU technology has vacillated wildly, and funding for Mars ISRU was essentially zero for the many years while funds poured in for lunar ISRU. This situation was changed remarkably in late 2013 when NASA announced an opportunity to land an ISRU prototype system on the Mars 2020 rover. This program will be reviewed in some detail in this book. Although this new sizable investment represented an important technological advance, it is not clear how this step fits into any presumed long-range, master plan.

In this book, I review the resources available for ISRU on the Moon and Mars, and the technologies that have been proposed for implementation. I also discuss how ISRU would be implemented within human mission scenarios, and I compare the missions with and without ISRU as well as can be done considering the limited available data. As one might expect, the most likely possibility for ISRU to become a viable benefit to a human mission is in providing ascent propellants for the return trip to Earth. While this makes good sense on Mars, unfortunately there are great difficulties on the Moon. None of the processes for producing oxygen from lunar regolith appear to be economically viable. The process for retrieving water ice from shaded lunar craters presents great difficulties. Further to add to these impediments, late lunar mission plans from the Griffin era called for use of space storable ascent propellants on the Moon, thus eliminating any demand for oxygen (produced by ISRU) as an ascent propellant. If that were not enough, safety considerations require that the Moon lander retains ascent propellants to allow for "abort to orbit" during descent in case of abnormal conditions. In which case, the lander would carry its own ascent propellants. Yet, NASA has spent millions of dollars over the past several years developing laboratory prototypes of arcane processes that produce oxygen for lunar ISRU that are very unlikely to be used.

Use of ISRU to produce ascent propellants on Mars appears to be viable and cost effective, but there are hurdles to be overcome. It appears certain that oxygen (and probably methane as well) will be used for ascent from Mars. This assures that propellants produced by ISRU on Mars are applicable to missions. There are two potential resources on Mars: CO_2 in the atmosphere and H_2O in the near-surface regolith. Two processes have been proposed for the utilization of only the CO_2 in the atmosphere to produce oxygen. Solid oxide electrolysis has been advanced to a fairly mature state with the implementation of the MOXIE Project. Alternatively, the so-called reverse water gas shift (RWGS) process may be worthwhile. Unfortunately, after funding an initial innovative RWGS breadboard study by Zubrin and co-workers that generated some optimistic results in 1997, NASA turned a cold shoulder on this technology and did not fund it for the next 15 years while they spent millions on impractical schemes for lunar ISRU.

A well-studied, practical Sabatier–Electrolysis process exists for producing CH_4 and O_2 from CO_2 and H_2. The problem for this process on Mars is obtaining hydrogen. Early NASA mission plans hypothesized bringing the hydrogen from Earth, but they seem to have underestimated the technical difficulty in doing this. Even more important is the fact that storing hydrogen on Mars is very difficult. There are indications of widespread deposits of near-surface H_2O on Mars, even in some near-equatorial regions. If this were accessible, it would provide an extensive source of hydrogen. Thus, the main problem for this form of ISRU on Mars is not process development, but rather, prospecting to locate best sources of near-surface H_2O.[3] What is needed is long-range, near-surface observations with a neutron spectrometer in the regions of Mars identified from orbit as endowed with near-surface H_2O. This might involve balloons, airplanes or gliders, network landers, or possibly an orbiter that dips down to low altitudes for brief periods. None of these technologies seem to be high on NASA's priority list.

The most immediate and feasible application of ISRU is to produce ascent propellants on Mars. Crew size determines the amount of ascent propellants needed and thus determines the leverage gained by using ISRU. Thus, crew size is one of the most critical factors in determining the feasibility and cost of a human mission to Mars.

Hence, I have concluded the following:

- None of the lunar ISRU technologies appear to be economically viable.
- The Mars RWGS process might be a viable option, but NASA's non-funding of this technology after an initial somewhat successful study leaves a great deal of uncertainty.

---

[3]We use the term H_2O rather than water here because it is not known whether the H_2O exists as water ice or mineral hydration.

- The MOXIE Project is demonstrating the viability and maturity of solid-state electrolysis for producing oxygen from Mars CO_2. Oxygen-only ISRU appears to be eminently practical.
- The Sabatier–Electrolysis process for Mars ISRU is technically and economically viable if a source of hydrogen can be provided. Bringing hydrogen from Earth and storing it on Mars is problematic. Prospecting for near-surface H_2O on Mars requires a costly campaign, yet the payoff is great.
- Depending on the availability of near-surface gypsum deposits at viable landing sites, utilizing near-surface H_2O on Mars might turn out to be the most cost-effective and technically practical way to utilize ISRU to enhance human missions in space.
- Crew size is one of the most critical factors in determining the feasibility and cost of a human mission to Mars. Crew size determines the leverage from using ISRU. The optimum crew size for a human mission to Mars has yet to be determined.

In the longer run, the optimal path for ISRU might possibly be to retrieve water from Near-Earth Objects and store this water in the cislunar region of space. This water would be convertible into hydrogen and oxygen propellants that could be used for departure from LEO and any other space mission transfers. To validate and implement such a scheme will require considerable dedication and up-front funding, neither of which seems to be readily forthcoming.

South Pasadena, USA   Donald Rapp

What's New in the Second Edition

1. Human Missions to Mars and Ascent from Mars Surface

In the first edition, I did not include a description of human missions to Mars, but rather, I simply referred to my book "Human Missions to Mars". However, as I show in this book, by far the most important application of *In Situ Resource Utilization* "ISRU" is to produce 30–40 metric tons of propellants for ascent from the Mars surface. To properly understand the value of Mars ISRU, one must see how ISRU fits into the bigger picture of the Mars mission. To that end, I now provide in this second edition a short synopsis of the Mars mission. In addition to that, I now provide for the first time, a detailed description of the process of ascent from Mars based on new NASA publications since the publication of the second edition of my Mars book. This, for the first time, clarifies the requirement for how much ascent propellants are needed. In doing this, I concluded that this critical figure is sensitively dependent on crew size. Therefore, I provided for the first time, a review and analysis of pros and cons for various crew sizes relevant to ISRU.

2. Solid Oxide Electrolysis of CO_2

In the first edition, the only information available was early, relatively trivial experiments on the laboratory bench in the 1990s. However, starting in 2014 (and continuing today) NASA funded a major new development to demonstrate this technology on Mars with a 2020 launch. That project is called "MOXIE". (see: https://mars.nasa.gov/mars2020/mission/instruments/moxie/).

I am a co-investigator on this project, so I have inside access to data and progress. I have included a lengthy section on solid oxide electrolysis that promises to be an important part of the Mars ISRU picture.

3. Ancillary Needs for Mars ISRU

    (a) It is not enough to produce propellants for ascent from Mars. The propellants produced in a chemical plant must be transferred to the Mars Ascent Vehicle and maintained cryogenically. I included a discussion of this based on recent NASA studies.

(b) Power: ISRU systems are power hungry. There has been new research on nuclear versus solar power systems on Mars, and I have summarized this relevant to ISRU.

4. Obtaining Water on Mars

If water can be obtained on Mars (not yet proven), that would have a major impact on the optimum approach for ISRU. I have upgraded this section based on new data and analysis by NASA.

5. Value of ISRU

I have totally revamped and rewritten this section, and I think it is now much more accurate and clear.

6. Recent NASA Plans

This section is new.

7. In addition, I went over every paragraph, sentence, and word and made a large number of minor changes for accuracy and clarity.

Contents

Nomenclature

AC	Ascent capsule
ALS	Advanced life support
ASR	Area specific resistance
CAC	Cryogenic accumulation chamber
COTS	Commercial off-the-shelf
DRA	Design Reference Architecture
DRM	Design Reference Mission
ECLSS	Environmental Control and Life Support System
EDL	Entry, descent, and landing
ERV	Earth return vehicle
ESAS	Exploration Systems Architecture Study
ESM	Equivalent system mass
EVA	Extra vehicle activity
FSP	Fission space power
HEO	Human Exploration and Operations
HEPA	High-efficiency particulate air
IMLEO	Initial mass in low Earth orbit
ISPP	In Situ Propellant Production
ISRU	In Situ Resource Utilization
ISS	International Space Station
JSC	NASA Johnson Space Center
LDRO	Lunar Distant Retrograde Orbit
LEO	Low Earth orbit
LL1	Lunar Lagrange Point #1
LLO	Low lunar orbit
LLT	LL1-to-LEO Tanker
LM	Landed mass
LMA	Lockheed Martin Astronautics
LOX	Liquid oxygen
LSS	Life support system

LWT	Lunar water tanker
MAAC	Mars atmosphere adsorption compressor
MAV	Mars Ascent Vehicle
MIP	Mars ISRU precursor
MMG	Mars mixture gas
MOXIE	Mars oxygen ISRU experiment
MSL	Mars Science Laboratory
NEO	Near-Earth Object
NERVA	Nuclear Engine for Rocket Vehicle Application
NTP	Nuclear thermal propulsion
NTR	Nuclear thermal rocket
RCS	Reaction control system
RESOLVE	Regolith and Environment Science and Oxygen and Lunar Volatile Extraction
RTG	Radioisotope thermoelectric generator
RWGS	Reverse water gas system
S/E	Sabatier/Electrolysis system
SEP	Solar electric propulsion
SH	Surface habitat
SOXE	Solid oxide electrolysis
SS	Space shuttle
ST	Space technology
TMH	Trans-Mars habitat
TMI	Trans-Mars injection
TPS	Thermal protection system
WAVAR	Water vapor adsorption reactor
YSZ	Yttria-stabilized zirconia

List of Figures

List of Tables

Mars ISRU

1

1.1 Human Missions to Mars

1.1.1 Background

Portree (2001) wrote a superb history of mission planning for sending humans to Mars as of year 2000.

According to Portree:

> More than 1000 piloted Mars mission studies were conducted inside and outside NASA between about 1950 and 2000. Many were the product of NASA and industry study teams, while others were the work of committed individuals or private organizations.

Starting with von Braun's vision of the 1950s, many attempts were made to define a feasible human mission to Mars. Development of nuclear thermal propulsion (NTP) began in the 1950s and 1960s. In 1968, Boeing published a fairly detailed design of a human mission to Mars making extensive use of NTP. The Boeing 1968 study set a high bar for extraordinary detail in analysis of the many subsystems and components needed for a human mission to Mars. The lengthy report is replete with detailed tables and illustrations. It is unfortunate that many subsequent studies failed to come close to clearing that bar. Mission concepts continued to appear through the 1970s and 1980s. In the 1990s, NASA produced *Design Reference Missions* DRM-1 and DRM-3 that became the original standard bearers for Mars mission planning. These DRMs introduced ISRU, and continued to rely on NTP. In the same time frame, Zubrin developed the *Mars Direct* concept and a group at Caltech developed the *Mars Society Mission* concept. In 2005, NASA published a summary of an extensive mission study known as DRA-5, but detailed backup for this summary does not seem to be available. In 2014, NASA announced the *Evolvable Mars Campaign* (EMC). However the EMC seems to be based on vague and ephemeral plans, with a lack of detailed engineering calculations. It will likely end up being scrapped

© Springer International Publishing AG 2018
D. Rapp, *Use of Extraterrestrial Resources for Human Space Missions to Moon or Mars*, Springer Praxis Books, https://doi.org/10.1007/978-3-319-72694-6_1

for good reasons, as NASA moves on to its next long-range plan. After 60+ years of planning, we still don't have a credible plan for a human mission to Mars. Meanwhile, Elon Musk continues to put out nonsensical dreams of Mars "colonization" to a gullible media in his periodic press releases.

Rapp (2015) amplified and updated Portree's history of planning for human missions to Mars. In this chapter, I will not reiterate that entire history, but instead I will proceed to more recent studies with particular emphasis on the role of In Situ *Resource Utilization* (ISRU).

One of the major factors that will determine the cost and logistic feasibility of a human mission to Mars is the number of heavy-lift launches of materiel will be required, as well as the complexity of logistics of on-orbit assembly, if needed. This is characterized in terms of the *Initial Mass in Low Earth Orbit* (IMLEO). Oddly enough, the way to estimate IMLEO begins at the end points of the mission:

1. How much mass must be lifted from the Mars surface to rendezvous in Mars orbit?
2. How much mass must be delivered from LEO to the Mars surface?
3. How much mass must be delivered from LEO to Mars orbit?
4. How much mass must be returned from Mars orbit to Earth orbit?
5. How much mass must be returned from Earth orbit to Earth?

The contribution to IMLEO from each of the above mass transfers is the mass transferred, multiplied by its appropriate "gear ratio". The total IMLEO is a sum of these contributions. A gear ratio is the ratio of the initial mass at the starting point to the final delivered mass for any transfer of payload. For example, if it requires say, 4 initial mass units to transfer one mass unit to its destination, the gear ratio for that transfer is four. My rough estimate for the gear ratio for M_1 is roughly 56. That is, it takes 56 mass units on the surface to deliver 1 mass unit of payload to rendezvous in elliptical Mars orbit.[1] The gear ratio M_2 for dry cargo is approximately 10. For crew, it is undoubtedly much larger. For cryogenic propellants it is anyone's guess. My wild guess is that to deliver 1 mass unit of cryogenic propellant from LEO to the Mars surface, the gear ratio is something like 16:1. The overall gear ratio from LEO to the surface of Mars, and from the Mars surface to Mars orbit is thus at least 500. It is evident that minimizing the total mass delivered to the Mars surface is important. Even more important is minimizing the mass of the ascent system from Mars to rendezvous in orbit since it has such a large gear ratio. That is why ISRU is so valuable, since it produces ascent propellants on Mars. The masses delivered to Mars and lifted from the Mars surface are strongly dependent on the crew size. That is why crew size is the most important single factor in determining the scale of a human mission to Mars.

---

[1]See paragraphs toward end of Sect. 1.1.4.

1.1.2 The Likely Mars Mission Scenario

Any mission to Mars is constrained by the following:

- The Earth and Mars go around the Sun at different speeds.
- The Mars year is roughly 22.57 Earth months. The Earth year is 12 months.

Starting at any given time when the Earth and Mars are juxtaposed in some relative position, we can fast-forward approximately 25.62 months.

- During this period of time, the Earth has made two transits around the Sun and is now 1.62 months further along its orbit than it was at the start.
- Mars made one transit around the Sun and is now 3.05 months further along its orbit.
- But 1.62 Earth months is about 48.6° along its orbit while 3.08 Mars months is 48.6° along its orbit.
- Thus the relative positions repeat themselves roughly every 25.6 months, but the actual positions have moved roughly 49° forward along the orbits.

Thus we conclude that the position of Mars relative to the Earth is at a propitious distance for transfer from Earth to Mars roughly once every 25.6 months.

If a spacecraft leaves Earth and takes say, 7 months to get to Mars, by the time it arrives at Mars, the Earth has moved halfway around the Sun and no return is feasible.

The Earth moves back into a position for transfer from Mars to Earth about 24 months after the initial transfer from Earth to Mars.

Thus a Mars mission time line might be something like:

- Month 1: Send spacecraft toward Mars
- Month 7: Arrive at Mars
- Months 8–23: Remain at Mars
- Month 24: Depart Mars for Earth
- Month 30: Return to Earth.

Thus the crew must remain on Mars for about 16 months. There is no way to abort the mission.

Returning from Mars imposes stringent requirements. The crew must be housed in a suitable habitat for about seven months, life support must be provided without interruption, and large amounts of propellant are needed for transfers. Liftoff from the surface of Mars requires a large mass of propellants, proportional to the liftoff mass. If the mission plan was to rise off the surface of Mars and go directly back to Earth, the entire mass of habitat plus life support would have to be lifted from Mars and sufficient impulse applied to send it on its way toward Earth. The amount of propellants required would exceed 100 metric tons (mT). The scale of the propulsion system for the return vehicle would be huge. The volume of the propellants would exceed 120 m^3, requiring more than 27 tanks of 2-m diameter. Instead, a sizable Earth Return Vehicle (ERV) is placed in Mars orbit containing the habitat for the seven-month journey back to Earth, along with the life support system. A very minimal capsule is used to transport the crew from the surface of Mars to rendezvous and transfer to the ERV in a couple of days. Even with this minimal capsule, the requirement for ascent propellants will be at least 30 mT.

The likely Mars mission strategy was laid down in the 1990s by a NASA study referred to as "DRM-1".

The strategy chosen for the DRM-1, generally known as a "split mission" strategy, breaks mission elements into pieces that can be launched from Earth with very large launch vehicles, preferably without rendezvous or assembly in low Earth orbit (LEO). The strategy has these pieces rendezvous on the surface of Mars, which will require both accurate landing and mobility of major elements on the surface to allow them to be connected or to be moved into close proximity. Another attribute of the split mission strategy is that it allows cargo to be sent to Mars without a crew at one or more opportunities prior to crew departure. This allows cargo to be transferred on low energy, longer transit time trajectories and the crew to be sent on a higher energy, shorter transit time trajectory. Dividing each mission into two launch windows allows much of the infrastructure to be emplaced and checked out before committing a crew to the mission, and also provides a robust capability, with duplicate launches on subsequent missions providing either backup for the earlier launches or growth of initial capability. Of supreme importance, the *Earth Return Vehicle* (ERV) will be emplaced in Mars orbit, and the propellant tanks of the *Mars Ascent Vehicle* (MAV) on the Mars surface will be filled (using ISRU), prior to launch of the crew.

In DRM-1, the first three launches will not involve a crew but will send infrastructure elements to low Mars orbit and the surface of Mars for later use (see Fig. 1.1).

Whether subsequent missions would go to the same location as the first mission, or to different locations on Mars is still being debated.

Three cargo vehicles depart Earth about 26 months before the astronauts, on minimum energy trajectories direct to Mars (that is, without assembly or fueling in LEO). After the

Fig. 1.1 Mission sequence for first mission to Mars in DRM-1

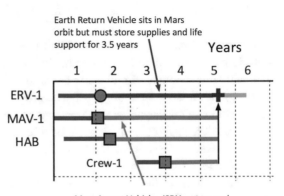

descent stage lands on the surface, the nuclear reactor autonomously deploys itself a considerable distance from the MAV. The MAV tanks are gradually filled with propellant prior to astronaut departure from Earth over a period of about 14 months.

The first crew of six astronauts departs for Mars (Launch 4) in the second launch opportunity. DRM-1 assumed that the capability of very precise landing on Mars can be developed technically, and that all assets for each flight can be integrated on Earth and simply joined on Mars.

The ISRU system fills the MAV tanks in about 14 months—prior to departure of astronauts from Earth. This stems from the facts that (i) launches are spaced at 26 month intervals, (ii) delivery of the MAV from Earth to Mars takes about 9 months, (iii) at least a month is allocated for set-up, and (iv) it is presumed that preparing for crew departure requires about 2 months. That is, the decision to send the crew will not be made until ISRU is complete. But once that decision is made, it will take at least 2 months to prepare for departure. Hence: $(26-9 - 1-2) = 14$. It will be noted that in several places in this book, I point out that NASA assumed 16 months for the duration of ISRU. I believe 14 months is more accurate.

The amount of propellant required for ascent depends on the orbit for rendezvous, the specific impulse of the ascent propulsion system, the mass of the ascent propulsion system (tanks, thrusters, plumbing, ...), and the mass of the fully loaded capsule used for transporting the crew to the ERV waiting in orbit. The mass of the ascent capsule is roughly proportional to the crew size. Later in this book, I discuss the issues associated with crew size, and how it affects the requirement for ascent propellants. This, in turn, has a major impact on the value of Mars ISRU. While DRM-1 and most other mission plans called for a crew of six, recent NASA studies were based on a crew of four. The requirement for ascent propellants for a crew of six is probably 50% greater than for a crew of four. The amount of propellant needed for ascent is the most crucial figure in regard to ISRU for a human mission to Mars.

Power on Mars would probably be supplied by a nuclear reactor or reactors. During the $\sim$14-month period during which the ISRU system is filling the tanks of the MAV, most of the power is used for ISRU. After the crew lands, most of the power is used for crew support, although the tanks of the MAV must continue to be refrigerated.

1.1.3 Crew Size

1.1.3.1 Introduction

In the context of this report, "crew size" refers to the crew that lands on Mars and eventually ascends to rendezvous with the orbiter. The crew size impacts almost every element of a human mission to Mars. The crew size (together with any crew assigned to remain on the orbiter) determines the scale of the trans-Mars habitat and the provisions for life support within that habitat. The crew size determines the scale of the Mars surface habitat, the EDL system, the power system, and the ascent system. The crew size determines the size of the ascent capsule, which in turn, determines the amount of ascent propellants required, and this determines the relative value of ISRU.

> Ultimately, the crew size is the most important parameter that defines the magnitude of a mission to Mars.

There is no universal agreement on the optimum crew size. There is some minimum crew needed for operation, maintenance and overhead, and this is probably somewhere between 2 and 3 crewmembers, depending on the level of automation. The amount of science and exploration that can be done on the mission then depends on the number of additional crewmembers above this minimum. A crew of six would provide for a considerable amount of science and exploration and at the same time, provide some redundancy and safety. A crew of four would provide a more easily implementable mission, but with more risk and less opportunity for science and exploration.

The crew of the orbiter is also important. If the orbiter is populated with a small crew, this crew will be in the small confined space for at least 2.5 years. The question arises: Is the orbiter populated, and if so, by how many crew? How would this orbiter crew survive physically and emotionally for 30+ months in a very confined space? We are only concerned here with the surface crew size, but the size of the crew in the orbiter is also an important contributor to total mission cost.

1.1.3.2 Review of Some Studies on Crew Size

The International Exploration of Mars, 4th Cosmic Study of the IAA, International Academy of Astronautics (IAA) said:

> The size and cost of a mission are directly driven by crew size.
>
> Cost considerations have tended to drive the number of crew in recent studies to the lower limits of what is believed by the study authors to be a safe and reasonable crew complement. There is much uncertainty as to what this minimum number really is. A high priority for human factors studies is to better establish minimum requirements for Mars mission crew size and the habitation amenities necessary for the long duration of these missions.
>
> Picking the optimum crew size for a given Mars exploration scenario involves a trade-off between increasing the number of crew members at the cost of a heavier, more expensive transportation system versus the benefits of accomplishing more work in the same time and having a more robust system. Because it will be expensive to transport additional personnel to Mars, it is clear that each crewmember should be capable of supporting several critical mission needs. A partial list of mission jobs might include: pilot, navigator, physician/dentist, electronic maintenance, mechanical maintenance, computer software maintenance, and life support systems maintenance. Other important specialist skills include mission commander, exobiologist, planetologist, and resource scientist. In addition, all crew members can be expected to participate in EVAs, drive the rovers, operate cameras and other instruments, drill core samples, set up science experiments, perform logistics functions, cook, clean, and

communicate with ground control personnel. Assuming that each crewman is trained in three specialist skill areas, and that a mission requires at least two crew trained in all critical skills and one crewmember trained in each of the other important specialist skills, the mission requirements imply a minimum crew size of six.

Landau and Longuski (2009) noted:

IMLEO is approximately proportional to the crew size, so the IMLEO may be scaled for an arbitrary crew size.

They used a crew of four in their mission model but suggested that the masses would scale as a factor of 1.5 for a crew of six. The crew size was probably chosen as four for expediency, yet as they pointed out, one can scale to the crew size. They also said:

We may wish to double the crew size (to eight) and [thereby] expand our capability to explore Mars.

Lamamy et al. (2005) said:

Concepts for the human exploration of Mars usually assume a crew of four to eight people. However, no study has provided a rationale for the composition and size of the crew based on a comprehensive set of factors. The suggested crew selections are instead dictated by the consideration of a single metric. On the one hand, advocates for a smaller crew of four are driven predominantly by a cost minimization. On the other hand, more conservative studies recommend a baseline crew size of six, citing safety as the primary concern.

They proposed to develop "a quantitative rationale for the crew design of missions to Mars based on a broader range of criteria". They discussed a number of factors:

Public visibility: This would be roughly independent of crew size unless a larger crew provided more groundbreaking discoveries.

Group dynamics: They quoted the study by Dudley-Rowley et al. (2002) that suggested

... crews of four or more containing men and women of different nationalities experience the fewest conflicts during a mission. However, experience from research ships and the Biosphere 2 experiment show teams of six or more tend to split into subgroups detracting from the overall cohesion of the team.

Skill mix: Analysis of the skills required led to the conclusion that a crew of four could "cover the skill areas and be operationally efficient". On the other hand, they pointed out the benefits of having additional scientists.

Value delivered: Lamamy et al. (2005) said:

The human exploration of Mars has several stakeholders encompassing the scientific and explorer communities, the economic and security sectors, and the general public. The interest of the general public is already captured in the public visibility metric. In a first phase, the analysis of value delivered focuses on the scientific and explorer stakeholders, who are most impacted by the crew selection.

With a crew of four, it is likely that about 80% of crew time would be used for self-maintenance and overhead, while roughly 20% of crew time would be available for science and exploration. With a crew of six, this ratio might change to 60–40. However, at this early stage, these figures are quite subjective.

Cost: Lamamy et al. (2005) quoted a study that estimated that addition of a crewmember would increase IMLEO by 74 mT. I think this is an underestimate, but it doesn't change their conclusion that mass-wise, a crewmember might be replaced by automation equipment.

Risk and Resiliency: Obviously, a larger crew is more resilient to an accident or problem with one of the crew.

Automation: Lamamy et al. (2005) concluded that use a crew of five plus 13 mT of automation equipment would return essentially the same value as a crew of six without the 13 mT automation payload. In their view, the gear ratio for transfer of cargo from LEO to the Mars surface is about six; therefore they concluded that there is an equal trade between a crew of six and a crew of five with automation. I think the gear ratio is at least ten but one crewmember is equivalent to a heavy lift launch; therefore I think the addition of 13 mT of cargo would add a heavy lift launch. Thus, I would agree with their conclusion that a crew of five plus automation might replace a crew of six with not much mass impact, although my numbers are quite different.

In summary, Lamamy et al. (2005) favored a crew of five with automation.

Salotti et al. (2014) pointed out that crew size for a human mission to Mars is typically between three and six astronauts, and considering human factors, a crew of three is possible but a crew of six is more appropriate. But Salotti et al. (2014) emphasized their analysis of the mission benefits of a crew of three. They estimated that IMLEO for the landed habitat for a crew of three would be about 2/3 of IMLEO for a crew of six. They also argued that with a crew of three, the EDL system could probably be greatly simplified. In addition, considering the total reduction of the payload that has to be sent to Mars, it might be possible to avoid LEO assembly and thus to send the interplanetary vehicles directly to Mars. They concluded:

> Reducing the size of the crew from six to three astronauts is probably a game-changing option. In addition, if six astronauts were the preferred crew size on the surface of Mars, the best option according to the IMLEO criterion would be, without question, a duplication of the scenario with three astronauts.

Salotti et al. (2014) acknowledged that a crew of six would be "better" than a crew of three, but they argued the difference would be quantitative rather than qualitative, and the price paid in mass and complexity with a crew of six is too high. I think their estimates of mission simplifications due to the smaller crew are probably reasonable, but I am not at all convinced that a crew of three is viable from the point of view of human factors, risk, and useful time available for science and exploration.

Dudley-Rowley et al. (2002) conducted a generic study of the effects of crew size on exploration missions using data from seven missions on Earth and three in space. In their "Pilot Study" they pointed out that the "common wisdom" was:

1. Heterogeneous crews with respect to nationality, sex, age, and experience will have higher rates of deviance and conflict than homogenous crews.
2. Larger crews will have higher rates of deviance and conflict than smaller crews.
3. Rates of deviance and conflict will increase with increasing mission duration.
4. The rate of deviance will peak in the third quarter (after the mid-point of the mission).

Dudley-Rowley et al. (2002) challenged these commonly held beliefs in their pilot study. They found major contradictions, in that heterogeneous crews and larger crews had lower rates of deviance and conflict, and furthermore, deviance and conflict tended to decline with increasing length of the mission. However, it is not clear to this writer how relevant all the data are to human missions to Mars.

They then carried out a secondary analysis in which seven factors emerged as important:

1. Increasing distance away from rescue in case of emergency (lessening chances of "returnability").
2. Increasing proximity to unknown or little understood phenomena (which could include increasing distance from Earth).
3. Increasing reliance on a limited contained environment (where a breach of environmental seals means death or where a fire inside could rapidly replace atmosphere with toxins).
4. Increasing difficulties in communication with ground or base.
5. Increasing reliance on a group of companions who come to comprise a micro-society as time, confinement, and distance leave the larger society behind, and where innovative norms may emerge in response to the new socio-physical environment.
6. Increasing autonomy from ground's or base's technological aid or advice.
7. Diminishing resources needed for life and the enjoyment of life.

Dudley-Rowley et al. (2002) concluded:

Crew size

Overall, larger crews were less dysfunctional than smaller ones. The crew which demonstrated the least deviance, conflict, and dysfunction of all was one that numbered about nine persons. These findings imply that habitat and workplace designers plan for larger facilities in extreme environments, such as those anticipated for Mars operations.

Heterogeneity

Any kind of heterogeneity (sex, ethnic, age, race, ...) studied herein seemed to be beneficial. Heterogeneity distinguishes people from one another, which makes them interesting to each another over the long haul and which offers complementary experiences and skills in order to allow the group to arrive at useful, innovative solutions during the expedition or mission.

Mission Duration

The longer missions in this study had lower rates of deviance, conflict, and dysfunction. Longer missions provide for a daily round of activities that resemble something more akin to an everyday, ordinary schedule. People have time to socialize and to get to know each other's strengths and weaknesses over a longer period.

Mission Interval

This study demonstrated dramatically that there are two distinctive patterns over mission intervals that are linked to heterogeneity and homogeneity of crews. More heterogeneous crews start out with a level of deviance, conflict, and dysfunction suggesting that they have some trouble coming to terms with their differences at the outset. However, they do come to terms with those differences and make use of the benefits of the heterogeneity that they possess. Their rates of dysfunction drop. More homogenous crews are different. Their deviance, conflict, and dysfunction rise sharply after the mid-point of the mission. They are likely working well together at the outset, as would be expected of people with similar training and background. However, their similarities become tiresome after a while.

Where none are available who have different life and work experiences, they may find themselves in situations where they cannot "think outside of the box" in a dilemma. The steep gradients between the third and fourth quarters of the missions of homogeneous crews suggest that they suffer much in attempting to reduce their dysfunction and achieve their mission objectives before end of mission.

1.1.3.3   Crew Size in Proposed Human Missions to Mars
Crew size is the most critical factor in determining the feasibility and cost of a human mission to Mars.

Rapp (2015) provided a history of proposals for human missions to Mars dating back about 70 years. Many such proposals have been made. One of the odd things about these proposals is the fact that relatively little discussion was given to crew size; yet crew size is the most critical factor in determining the feasibility and cost of a human mission to Mars. Given that we don't really know the ideal crew size, it is not unreasonable that a team might arbitrarily select a crew size in planning a human mission to Mars; yet it would be incumbent on them to estimate how their mission might scale to other crew sizes. Not much seems to have been done in this area.

A 1989 Soviet Plan for a manned Mars mission said:

A crew of six was determined to be the optimum size required to conduct a manned Mars mission. Although the crew size could be reduced, it is unlikely that a mission would be conducted with fewer than five members. A crew probably would consist of at least a commander, a pilot, a flight engineer and two mission specialists, one of whom might be a physician. We believe that at least three crewmembers would go to and from the surface of

Mars in an excursion module while the other two to three crewmembers would remain in orbit around Mars in a mission module.

Zubrin and Weaver (1993) provided interesting insights:

Much has been written about the desirability from a human factors point of view of having a large crew on a long duration Mars mission. However as the size of the crew drives the mass of all habitats, transportation stages and launch vehicles, it is essential from the point of view of cost and technical feasibility to keep the size of the crew to the absolute minimum. Furthermore, no matter how many back up plans and abort options are included in the mission design, it must be understood that in sending a crew to Mars we are definitely sending a group of humans into harm's way, and thus from a moral point of view the fewer people we have on board the initial missions, the better. Finally, no matter how desirable a large social group may appear to be for company on a long trip, any examination of the history of human exploration on Earth will show that it is entirely possible for long duration expeditions to be carried out successfully by one person, two people, or any other number.

The question then, is how many people are really needed on a piloted Mars mission. If the mission were to fail, far and away the most likely cause would be failure of one or more of the mission critical mechanical and electrical (propulsion, control, life support) systems employed. The most important member of the crew then, the one on whom all other depend for their lives, is the mechanic. Call this person a flight engineer if you wish, (he or she is an engineer in the sense of an old-time railroad locomotive or steamship engineer) but what is needed is an ace mechanic who can sniff out problems before they occur and fix anything that can be fixed. This job is so critical that, despite our preference for small crews, we recommend carrying two people capable of handling it. The next most important job on the mission is that of field scientist The reason for this is simply that the purpose of the mission is to explore Mars, and so after those needed to get the crew to Mars and back, the next most important personnel are those essential to competently carrying out the exploration goals of the mission. Since no science return would also effectively be a form of mission failure, once again we recommend carrying two people that can do this job. One of the field scientists should be a geologist, oriented towards exploring the resources and understanding the geologic history of Mars. The other should be a biogeochemist, oriented towards exploring those aspects of Mars upon which hinge the question of past or present life. The biogeochemist would also conduct experiments to determine the chemical and biological toxicity of Martian substances to terrestrial plants and animals, and the suitability of local soils to support greenhouse agriculture. And that's it. There is no need for people whose dedicated function is "mission commanders," "pilot," or "doctor." The mission will need someone who is in command, and a second in command for that matter, because in dangerous circumstances it is necessary to have someone who can make quick decisions for all without the need for electioneering or debate. But there is no room for someone whose sole function is to manage others to get the job done. Similarly there can be no one on board whose job description is "pilot." The spacecraft will be capable of fully automated landing, and piloting skills would at most be useful as a contingency backup to the automated flight system during a few minutes of the 2.5 year mission. If such a manual flight control backup is desired, then one or more members of the crew could be given cross training as a pilot. (It is much easier to train a geologist to be a pilot than a pilot to be a geologist.) Finally, no ship's doctor as such. (It may be noted that the great Norwegian explorer Amundsen always refused to carry doctors on his expeditions, noting that they were injurious to morale and that the large majority of medical emergencies that occur on expeditions can be handled just as well by experienced explorers.) All members of the crew will be trained in first-aid, and expert systems on board and medical consultation from Earth will be available to diagnose readily treatable conditions (i.e. ear infections and the like). Such diagnoses could be assisted by having a member of the crew be someone who had either practiced general medicine at some point earlier in his or her career or who had been crossed trained to the level of knowledge of a medical assistant, and equip either with a country doctor's black bag. The biogeochemist would be a natural candidate for such a cross-trained role. However the idea of having a dedicated top-notch doctor on-board who spends his or her time on the mission reading medical texts and honing skills by practicing surgery with virtual reality gear, or worse, making a pest of himself by subjecting the rest of the crew to an in-depth ongoing medical study, is simply ludicrous. To summarize in "Star Trek" terminology, what a piloted Mars mission needs are two "Scotty's" and two "Spock's". No "Kirk's," "Sulu's," or "McCoy's" are needed, and more importantly neither are the berths and rations to accommodate them.

As always, Zubrin's observations are interesting and insightful, but at the same time controversial and biased toward optimistic minimization.

Hirata et al. (1999) developed a detailed mission plan for a human mission to Mars. They said:

A crew of five was determined … based on the minimum of four for adequate science return and system maintenance advocated by Mars Direct, with the addition of a crewmember for medical duties as advocated by the DRM-1. Instead of sending two medical crewmembers as in DRM-1, however, at least one of the science crew will be able to supply medical treatment in the event of the medical officer's illness or injury. Five crewmembers were thus determined to be sufficient for science return, maintenance of systems, and medical upkeep of the crew …. The rapid accumulation of habitats and other infrastructure at a single point on the Martian surface … for later missions could allow for greater crew size on succeeding missions.

The NASA Design Reference Mission DRM-1 said:

The nominal crew size for this mission is six. This number is believed to be reasonable from the point of view of past studies and experience and is a starting point for study. Considerable effort will be required to determine absolute requirements for crew size and composition. This determination will have to consider the tasks required of the crew, safety and risk considerations, and the dynamics of an international crew. Crew members should be selected in part based on their ability to relate their experiences back to Earth in an articulate and interesting manner, and they should be given enough free time to appreciate the experience and the opportunity to be the first explorers of another planet. Significant crew training will be required to ensure that the crew remains productive throughout the mission.

Hornbeck et al. (2003) described ESA plans for human exploration. In dealing with a human mission to Mars, they said:

Crew education should focus on scientific (e.g., biology, medicine, psychology, geology, atmosphere, meteorology, and astronomy) as well as technological skills (commander, spacecraft engineering, manufacturing technology, navigation, communication, software engineering). Assuming that each crewmember is intensively educated in one scientific and one technological area, this leads to a crew size of six.

No empirical data from spaceflight, analogue environments or simulation studies are available to define optimal crew sizes for long-duration missions under conditions of isolation and confinement. Whereas a small crew size can lead to detrimental effects of social monotony, a large crew increases the risk of clique formation, which has been found to impair crew performance. From a psychological point of view, though, it appears that a larger crew, consisting of six to eight individuals, would be better for crews living on the lunar or Martian surface, in order to counter the social monotony. A particular feature of the 1000-day Mars reference mission is the two-person crew which is planned to be left in Martian orbit in order to monitor and maintain the technical systems of the orbiter. Beside all other problems associated with this feature (e.g. motivation, sensory deprivation due to environmental monotony, boredom due to restricted range of tasks), it is highly likely that social monotony will lead to severe problems for interpersonal interactions and co-operation, as well as for their individual behavioral health and performance. Early simulation studies suggest that dyads fare even worse than triads under conditions of isolation and confinement (i.e. they show higher levels of anxiety and express more complaints), although triads are usually considered to be more difficult with regard to group dynamics. Even though the Mir station usually had a permanent crew of two, the effects of social monotony were effectively countered by the presence of visiting crews and intensive support from ground. This cannot, therefore, be taken as a positive example for a mission to Mars. Whenever possible, a two person crew should be avoided.

This report went on to discuss many aspects of crew selection from various points of view. One of these related to gender issues, as follows:

Gender issues have to be carefully considered in selecting crews for a long-duration exploratory mission. This holds, in particular, for both Mars mission scenarios, because of their extraordinary length and the high levels of confinement, isolation, and crew autonomy involved. No systematic research has ever addressed behavior and performance issues arising under confinement and isolation in crews with the same and mixed sexes. There is virtually no experience with all-female crews and research with mixed-gender crews in spaceflight, simulations, and analog environments have provided inconsistent results. Heterosexual activity within mixed-gender crews is usually considered to be more conducive to distur- bances in intra-crew relations. Risks commonly associated with sexual activity between crewmembers during long-duration missions include:

- disturbances of crew-cohesiveness by the formation of couples
- interpersonal tensions due to jealousy
- sexual deprivation can be harder to endure in the presence of persons perceived as sexually attractive
- sexual harassment and violation
- unintentional pregnancy.

Only the last of these risks is strictly associated with heterosexual activity, whereas the others can be associated with homosexual activity as well. Furthermore, the first two also occur if strong friendships develop between certain crewmembers (of the same or different sex) without any sexual content. Specifically in order to address problems of sexual depri- vation, it has been proposed to send only successfully married couples on a mission to Mars. However, this option does not appear to be realistic, given the specific profile of professional skills in such crews. Despite the fact that this solution might soften moral considerations, it neither presents an effective countermeasure for risks like unintentional pregnancy, nor for disturbances of crew-cohesiveness. The same holds for the suggestion to define a clear code of conduct for a mission (including rules concerning sexual activity). Such rules may easily lose any force, given the high crew autonomy and lack of external control, particularly during a Mars mission. In any case, sending mixed-gender crews with females of reproductive age on a long-duration mission to the Moon and Mars will require effective means of contraception which can be provided to male and/or female crew members and which are suitable for use under the specific conditions of these missions (e.g., concerning pharmaco-dynamic and pharmacokinetic changes of contraceptive medication in different gravity environments). In addition, crews should include at least two members of the same sex in order to avoid problems of minority. Based on currently available data and anecdotal information, and given the inevitable issues and specific risk of pregnancy associated with sexual activity in mixed gender crews, it is clear that sending mixed gender crews to Mars can be considered only if the issues mentioned above are dealt with openly, and proper countermeasures are made available to alleviate risks. In future, any discussion on sexual behavior in space should not be limited to heterosexual behavior alone. More research on this issue is definitely needed before any final recommendation can be given.

I regard this ESA Report as one of the best for dealing forthrightly with issues crew size and crew selection.

DRM-3 retained the crew size at 6. DRA-5 was not very specific on crew size but the implication seemed to be retention of a crew of six.

Manzey (2004) pointed out that a human mission to Mars "will provide unique psychological challenges that do not compare to any other endeavor humans ever have attempted". He said:

Missions to Mars will add a new dimension to the history of human spaceflight with regard to the distance to travel and mission duration. Accordingly it might be called into question to what extent our psychological knowledge derived from research during orbital spaceflight and in analog environments really can be applied to crews traveling through outer space.

Manzey (2004) discussed new psychological challenges in human missions to Mars: (1) individual behavior and performance, (2) crew interactions, (3) psychological countermeasures, and (4) Indeterminable psychological risks of Mars missions: the Earth-out-of-view phenomenon. While he did not overtly discuss the pros and cons of various crew sizes, he assumed a crew size of six.

Hofstetter et al. (2006) more or less summarized the issue of Mars crew size: *The expedient choice for ease in implementation is four. The desirable choice for a successful mission is six.*

Anna et al. (2012) proposed a human mission to Mars. Their comment on crew size was:

One solution for isolation problem is to increase number of crewmembers in order to increase human interaction. Four crewmembers is the smallest permissible crew size in the specification, partially in response to the isolation problem.

Salotti et al. (2012, 2014) defined an approach for a human mission to Mars. They pointed out that the original selection of a crew of six in NASA DRMs was somewhat arbitrary. They presented arguments for a crew of four:

Even if the skills of the astronauts enable a good exploitation of the spacecraft, we also have to consider scientific skills to perform experiments in space and on the surface of Mars. The 4 astronauts of our scenario can work together on the surface of Mars but in comparison with a crew of 6 with different specialists, there would be a lack of expertise in some domains with a possible negative impact on scientific returns. However, if that drawback enables strong simplifications of the scenario, which in turn would make manned missions less risky and affordable, is that not the most suitable option?

But their specific plan for a Mars mission used three vehicles with a crew of two per vehicle. Two vehicles landed so the surface crew was four, while two remained in orbit. The emphasis was on making the mission more affordable even if some accomplishments were sacrificed.

Moss (2013) developed a plan for human mission to Mars. He pointed out that the (Zubrin) Mars Semi-Direct and NASA DRA architectures specified crews of six, while the (Zubrin) Mars Direct and MARS-OZ architectures required only four. He chose a crew of six for several reasons, two of which were:

It's about as large a crew as we can handle while keeping the mission achievable and affordable.

A crew of six permits a dedicated crewmember for the most crucial functions, while also allowing for a degree of redundancy in skill coverage.

Genta (2015) summarized his view:

The number of people participating to the mission is one of the most important choices. The cost, the size and the complexity of the mission are strictly linked with the number of people on board and, at least for the first missions, this number should be kept to a minimum. However, there is no agreement on what this means. At least two aspects must be taken into account: the crew must include specialists in the many fields involved and the number must be large enough to avoid psychological problems. The astronauts can have an extended training period to reach the required competencies and many tasks can be automated. Each crewmember will be a specialist in at least two disciplines and mission control will supply a support in all specific fields and particularly in the management of emergencies, even if the crew must be able to face emergency situations on their own for at least the time required to receive suggestions and instructions from Earth. An acceptable minimum crew size from this viewpoint is often assumed to be 3. From the viewpoint of psychological factors a higher number, likely 5 or 6, may be required, although this number may be decreased by a suitable choice of the candidate astronauts and by a suitable psychological training. Moreover, in an international mission there will be pressures to include people from different origins and it is predictable that each partner will try to include its representative in the first mission. It is unlikely that a mission based on chemical propulsion, implying aerocapture in Mars orbit, is possible with a crew of 4 to 6, which would make nuclear propulsion mandatory. An alternative is splitting the crew into two groups, using 2 separate vehicles, each carrying 3 people, so that aerobraking is possible. This has also some beneficial effects on safety and reliability.

1.1.3.4   Psychological Aspects

Collins (1985) provided an early review of psychological aspects of astronauts in space. He mentioned:

> Psychological compatibility has been a recurring problem during short-duration missions…
>   A plethora of individual and interpersonal psychological problems have been documented, during and after space missions.
>   Poor judgment, belligerence, interpersonal dissension, irritability with ground managers, and gross violations of crew discipline.

Collins (1985) documented quite a number of specific aberrant behaviors in both US and Soviet space experience prior to 1985, and pointed out:

> Astronaut Schweichart experienced the interpersonal conflict aboard Apollo 9 and correctly predicted that as future missions get longer and the crews become larger, more intense interpersonal hostilities would occur.

He also reported on problems in confined isolation in extended underwater submarine cruises, as well as Arctic and Antarctic programs. The crew experienced the following:

> In the initial period all individuals experienced a sharp increase in anxiety, irritability, and difficulty sleeping. Work activity considerably reduced these unpleasant experiences. During the second and longest phase, anxiety diminished but feelings of depression increased. Finally, just before leaving the third (and last) phase, emotional expression increased and became less inhibited. Sleeplessness, a universal symptom associated with Antarctic isolation, was observed in all three phases.

Collins distinguished between *isolation* and *confinement*. Isolation implies removal from one's usual physical and social surroundings. Confinement refers to restriction within a portion of their environment.

Collins raised the questions:

> What sort of human can endure the utter isolation and severe confinement of a long space voyage? What sort of crew can make the voyage a successful one?
>   Most programs selecting personnel for work in isolated and confined backgrounds have emphasized the background, training, personality, interest, and aptitude characteristics of individual applicants or volunteers. However, in the case of small groups such as astronaut crews, where the interaction and interdependencies demand cooperative functioning and team orientation, … a selection program directed toward identifying the most effective "crews," rather than merely "qualified individuals," would seem to be a logical approach.

He then discussed crew size. By 1985, most US space missions used crews of 2 or 3. He concluded:

> The optimal planning of future manned space flights requires a better understanding of the effects that increasing group size will have on interpersonal dynamics. Substantially larger crews than dyads and triads have been forecast for future space flights. A thorough understanding of the effects that crew size will have on crew member performance and satisfaction awaits the results of carefully controlled studies of different sized groups operating under space flight-like conditions.

Stuster (2000) reviewed long-duration expeditions of the 19th century, and summarized psychological stressors that occurred in these expeditions.

Morphew (2001) listed stressors of long duration spaceflight as shown in Table 1.1. However he did not deal directly with the issue of crew size.

Sandal (2001) said:

> A possible mission to Mars, for example would present obvious technical problems; no less challenging, however, would be the complex issues raised by crew design. Considerable effort will be required to determine requirements for crew size and composition.

He emphasized the pitfalls in relating human factors in Earth analogs to long duration space missions.

Palinkas (2001) reviewed various psychological issues associated with long-term missions in space. Anecdotal reports of previous space flights cited "prevalence and severity of symptoms of depression, insomnia, irritability/anger, anxiety, fatigue, and decrements in cognitive performance". He discussed many aspects of the social dynamics of small groups in isolated and confined environments.

Wickman (2006) discussed physiological and psychological effects and counter-measures in a human mission to Mars. Stressors include:

- Isolation
- Confinement
- Limited habitation volume
- Compromised quality/conditions of habitation environment
- Absence of fresh air
- Reduced sensory stimulation
- Boredom
- Regimented work/rest schedules
- Strangeness of environment
- Awareness of risk.

Of particular interest was an extended discussion of habitat volume considerations. However, the paper did not seem to distinguish between volume considerations for the transit vehicle and the surface habitat. Nevertheless, the estimated need for volume per crewmember was given as (m^3):

Command/control	1.1
Payload/science	2.5
Kitchen/wardroom	1.5
Private hygiene	2.5
Sleeping quarters	1.2
Exercise area	3.0
Health/medical area	1.1
Total	12.0

Table 1.1 Stressors of long duration spaceflight

Physiological/physical	Psychological	Psychosocial	Human factors	Habitability
Radiation	Isolation and confinement	High team coordination demands	High and low levels of workload	Limited hygiene
Altered circadian rhythms	High-risk conditions and potential for loss of life	Family life disruption Limited equipment, facilities and supplies		Limited sleep facilities
Decrease in exposure to sunlight	System and mission complexity	Enforced interpersonal contact	Mission danger and risk associated with: equipment failure, malfunction, or damage	Lighting and illumination
Adaptation to micro-gravity	Hostile external environment	Crew factors (i.e., gender, size, personality, etc.)	Adaptation to the artificially engineered environment	Lack of privacy
Sensory/perceptual deprivation of varied natural sources	Alterations in sensory stimuli	Multicultural issues	Food restrictions/limitations	Isolation from support systems
Sleep disturbance	Disruptions in sleep (readjustment with crew changeovers) "Host-Guest" phenomenon	Technology-interface challenges		
Space adaptation sickness (SAS)	Limited habitability (e.g., limited hygiene)	Social conflict use of equipment in microgravity conditions		

Vakoch (2011) is a 200+ page compendium of discussions of essentially every psychological issue of humans in space missions. It is impossible to summarize all the many reports of studies in this document. Only two quotes specifically relevant to Mars missions are given here:

> The impact of isolation and confinement has been shown to be significantly impacted by various moderator variables, e.g., the difficulty of rescue. While an emergency on the International Space Station certainly poses difficulties regarding time to rescue, one can argue that the difficulties inherent in a Mars mission or even here on Earth from the Antarctic in midwinter, where weather conditions may absolutely make rescue impossible for long periods, carry a qualitatively different psychological impact. An emergency on a mission to Mars will preclude any chance of rescue and necessitate a high degree of autonomy for the crew in making decisions without any real-time mission support. The degree to which such factors magnify the negative effects of isolation and confinement is critical to assess.
>
> In the context of long-duration spaceflight such as a future Mars mission, there is still very little empirical evidence to inform on the ideal crew mix despite the many discussions that have ensued.

Kanas (2014) reported on

> … two NASA-funded international studies of psychological and interpersonal issues during a series of on-orbit missions lasting 4–7 months to the Mir and the International Space Stations. The Mir study sample consisted of 5 American astronauts, 8 Russian cosmonauts, and 42 American and 16 Russian Mission Control personnel. The ISS study sample consisted of 8 American astronauts, 9 Russian cosmonauts, and 108 American and 20 Russian Mission Control personnel. Subjects completed a weekly questionnaire that included items from a number of well-known measures that assessed mood and group dynamics. Both studies had similar findings, so in a sense they replicated each other. In both studies, there were no significant changes in levels of emotion and group interpersonal climate over time. Specifically, there was no evidence for a general worsening of mood and cohesion after the halfway point of the missions, and no evidence for a third quarter phenomenon. It should be noted that some individual crewmembers showed evidence for such a decrement just after the halfway point of their missions, but others showed no such effect or even experienced an improvement in emotional state during the second half. Our belief is that the absence of general negative time effects in our studies was the result of supportive actions taken by flight surgeons and psychologists in Mission Control, which included the sending of favorite food and surprise presents on resupply ships and increased communication with family and friends on Earth. Such actions helped to provide novelty and counter the effects of isolation, loneliness, and limited social contact. The celebration of mission milestones and holidays likely contributed to the maintenance of morale as well.

Kanas (2014) emphasized the importance of an identified leader, as well as frequent communications between crewmembers and people on Earth. He also discussed specific stressors appropriate to a human mission to Mars but was similarly upbeat, suggesting that these stressors could be overcome by careful crew selection and training. He did not comment on crew size.

My personal view is that a mission to Mars, lacking the ability to return home in a few days in case of emergency as in lunar missions, and requiring about two and a half years (or more) of close confinement in a dangerous shell that would mean death if penetrated, imposes a huge psychological stress far beyond any space induced stress imposed by orbital or lunar missions. The lack of privacy and likely primitive toilet facilities, especially in zero-g would add to the stress. Add to that the physiological impacts of radiation and low gravity.

The gender makeup of the crew is interesting. Sex remains the big unknown, silent topic. Is the crew committed to no sex for 2.5 years?

Thus the question of crew makeup in addition to crew size remains very uncertain. Yet crew size makes a huge difference for the ascent vehicle, and the size of the ascent vehicle determines the importance of Mars ISRU.

NASA's current approach for crew transport is the "Orion" transfer vehicle. Data for the Crew Module in Orion were given by NASA as:

$$\text{Crew capacity: } 2-6$$
$$\text{Dimensions: } 3.3 \times 5 \, \text{m}$$
$$\text{Volume (Pressurized): } 9.56 \, \text{m}^3$$
$$\text{Habitable Volume: } 8.95 \, \text{m}^3.$$

This seems appropriate for delivery of crew from Earth to LEO, but the habitable volume seems far too small for long transfers to Mars. In short, NASA does not have a vehicle to transport crew from LEO to Mars.

1.1.3.5   Earth Simulations and Analogs

There have been numerous Earth analog simulations of confinement in space missions, of various durations and with various constraints. I will only mention some of these briefly since I think that while they do provide insights into issues associated with confinement, they lack the factor of impending danger, which is the most vital aspect of the whole situation.

Clancey (2004) reported on a 12-day closed simulation in the Mars Desert Research Station in Utah, isolated from other people, while exploring the area and sharing daily chore. While 12 days is hardly enough to generalize, he made some astute observations of the needs for a crew and an appropriate habitat:

1. Survivability: Engineers are necessarily first concerned about physiological requirements of keeping the astronauts alive, with strong mass and cost constraints.
2. Comfort: The first contributions by human factors (industrial engineers) is to move beyond safety to improve comfort, with issues ranging from personal hygiene, privacy, and convenient "anthropometric" tools and designs...
3. Performance: The next level of human factors concerns task support or productivity, relating crew size, skills, tools, automation, procedure manuals, scheduling, training, computer interfaces, facilities layout, etc. These considerations may be matrixed against physiological constraints such as fatigue and affects of microgravity. More

broadly, industrial engineering observes and analyzes processes systemically to reduce cost and increase productivity.

4. Adaptability: Next, contributions by social scientists focus on crew teamwork, project collaboration (e.g., with scientists on earth), creative workarounds, informal assistance, replanning, and learning during the mission.

Crucially, a holistic perspective emphasizes that properties (e.g., "human capability") do not exist in isolation, but are relational, depending on context, which is always physical and interpersonal, as well as historical, involving both experience and expectation. Thus predicting what people can actually accomplish during a spaceflight, including their limitations and strengths, needs to be somehow triangulated from laboratory and workplace studies, similar space missions (e.g., relating Apollo to Mars), and mission simulations. As we move from the laboratory to analog missions on Earth to, for example, space station simulations of Earth-to-Mars transit, we will discover what combinations of facilities, roles, tools, etc. work in practice and gain increasing confidence about the right design for an actual mission. However, our theoretical stance tells us that even then we will have a lot to learn, and should design habitats and all aspects of the mission for adaptation during the mission.

NASA (2011) listed the various NASA programs in analogs with some description of the locales and the installations.

Sadler (2014) provided descriptions many of these analogs, but did not present any results.

Schlacht et al. (2016) provided information on a number of analogs, past, present and future, but did not report on what was learned from these studies.

It is difficult to find any references that summarize what we have learned from the analogs.

1.1.3.6 Habitats

It is important to realize that a human mission to Mars requires several Habitats to house the crew. Clearly, the size and associated equipment (e.g. life support, power, thermal control, recycling, …) will scale with crew size. The following are needed:

Small capsule for lift-off from Earth and transfer to trans-Mars habitat. Orion should suffice.

Trans-Mars Habitat (TMH). The requirements for the TMH depend on the crew size and mission scenario. The TMH will transport the total crew (surface crew plus orbiter crew) to Mars orbit over a period of six to nine months. The surface crew will only spend 12–18 months in the TMH for transport to and from Mars while the orbiter crew will spend an additional ~ 18 months in Mars orbit. Assuming the surface crew is chosen as 6 and the orbiter crew is 2, the total number of crew months supported by the orbiter is about $8 \times 15 + 8 \times 15 + 2 \times 18 \sim 276$ crew-months. These are all exposed to space radiation and zero-g.

Ascent Capsule (AC): The AC must carry the surface crew to orbit and perform a rendezvous with the orbiter, probably within two days. Although this is far less extensive in size and resources than the TMH or SH, its mass is multiplied by a gear ratio (from LEO) of about 56.

Surface Habitat (SH). The SH must provide for about $6 \times 18 = 108$ crew-months. These include ~ 0.3 g, and Mars surface radiation levels (lower than orbit levels).

1.1.3.7   Summary and Implications for ISRU

It seems apparent that the mass requirement (and cost) of a human mission to Mars roughly scale with surface crew size. The surface crew size is the single most important parameter in designing a human mission to Mars, yet as I have shown, relatively few studies have been carried out, and considerable further work is needed.

The requirements for surface crew size for a human mission to Mars include the need to provide important skills, provide protection for unforeseen risks via redundancy and medical expertise, as well as the psychosocial needs for a well functioning cohesive crew.

There is considerable agreement that a surface crew of six would be ideal although the basis for this is not yet very firm. However, mission models indicate that a surface crew of six would drive up the mass (and cost) to levels that would be difficult to fund. Therefore recent NASA mission planners have chosen a surface crew as small as four, thus making the mission more affordable, while sacrificing capabilities and adding risk. The exact details of the trade between 6 and 4 surface crewmembers in terms of cost, capabilities and risk remain as yet, not very clear.

The Evolvable Mars Campaign is based on use of a crew of four for lunar and asteroid missions as steppingstones to an eventual Mars mission. The current NASA assumption seems to be for a crew of four. However, until we better understand the trade between 4 and 6 surface crewmembers, such a decision for a crew of four seems to be premature.

An estimate of the requirement for ascent propellants for a crew of four was provided by Polsgrove et al. (2015) as 22.7 mT of oxygen and 7.0 mT of methane (see next section). This was based on a modeled 3.8 mT fully loaded ascent capsule with crew, samples and life support equipment. Experience teaches us that such early forecasts of subsystem masses tend to be low, and one might expect significant mass growth as designs mature further. With a crew of six, these propellant estimates would probably rise by somewhere between 40 and 50%, assuming some economies of scale. The full-scale requirement for ascent propellants might then ultimately turn out to be somewhere between the 30 mT estimated by Polsgrove et al. (2015) for a crew of four, and as much as 45 mT for a crew of six with mass growth in the ascent capsule. It would appear that ISRU for ascent propellants is a mission-enabling technology.

The ascent propellants in a human mission to Mars would be produced by ISRU over a time period prior to human departure from Earth. That time period is 26 months, less the time required to deliver the ISRU system to Mars and get it operating, less any time allocated to make a final decision to send the human crew, as well as time for final preparations once that decision is made. Delivery of cargo is likely to require 9 months, and at least 1 month is required to set up the power system and initiate the ISRU system. I assume that a 2-month period of grace for final decision and preparation would be needed, allowing up to 14 months for ISRU. Polsgrove et al. (2015) assumed 16 months of ISRU operations. Using Polsgrove's estimates for ascent propellants for a crew of four, and a 16 month ISRU duration, the full-scale oxygen production rate is 22,700/ $(16 \times 30 \times 24) = 2.0$ kg/h. With a crew of six, and a 14-month duration, the full-scale oxygen production rate is roughly estimated to be $31,800/(14 \times 30 \times 24) = 3.2$ kg/h, assuming no mass growth in the ascent capsule except for scaling. With mass growth, this could be higher.

In addition to mass considerations, volume of propellants is also important. For a crew of four, if 7 mT of methane and 23 mT of oxygen were transported from Earth to Mars it would require 37 m^3 of internal volume (nine 2-m diameter tanks plus insulation). For a crew of six, about thirteen 2-m diameter tanks would be needed. These would have to be maintained at cryogenic temperatures behind an aeroshell.

1.1.4   The Mars Ascent Vehicle

1.1.4.1   Introduction

A primary purpose of Mars ISRU is to generate some, or all of the propellants needed for ascent from Mars for rendezvous with the ERV waiting in Mars orbit.

Basically, there are two potential orbits that could be used for the ERV: a circular orbit or an elongated elliptical orbit. From the point of view of the MAV, the circular orbit would be preferable because the Δv required to ascend from the Mars surface to a circular orbit is roughly 4.3 km/s, whereas the Δv required to ascend from the Mars surface to an elongated elliptical orbit is about 5.6 km/s. However, the propellant requirement for capturing the ERV into Mars orbit as it approaches Mars is greater for the circular orbit than it is for an elliptical orbit. But if the majority of ascent propellants are produced on Mars by ISRU, it may be worth paying the price of using more ascent propellants, for the saving in propellants to put the ERV into Mars orbit (and the saving in the ERV departing Mars orbit). Therefore we shall assume here that the elliptical orbit will be chosen.

1.1.4.2   Ascent from Mars: MAV and Crew Capsule

There have been many studies of human missions to Mars over the years, but the details on the Earth Return Vehicle (ERV) orbit, the ascent trajectory, the ascent capsule, and the ascent propulsion system were sketchy at best. Assumptions for the specific impulse of the ascent propulsion system were typically overly optimistic.

Finally, in 2015, Polsgrove et al. (2015) (to be abbreviated "P15") presented an impressive conceptual design of a MAV with many details on the ascent propulsion system (although the presentation is rather terse on most other aspects of ascent). This paper is important because it fills what would otherwise be characterized as a vacuum regarding details of the MAV. It is also worth mentioning that they used a more credible specific impulse for $CH_4 + O_2$ propulsion of 360 s, whereas earlier NASA studies used unreasonably higher values.

The MAV consists of the crew capsule, the two-stage propulsion system, and supporting subsystems. As P15 put it:

> Crew ingress [into the capsule] is facilitated by a pressurized tunnel connected to the rover. The tunnel minimizes the potential to contaminate the MAV with Martian soil which would become airborne once the MAV reached orbit. It also allows the crew to leave surface exploration suits behind in the rover and enter the MAV and ascend in the lighter and more flexible launch and entry suits. One key assumption is that the crew [would] remain in their suits with visors up during ascent through rendezvous and dock except for brief hygiene breaks. This minimizes the "feed the leak" duration and consumables penalty in the event of a cabin leak and minimizes required cabin size by eliminating the need to store four empty suits.

The system for disengaging the tunnel from the MAV prior to liftoff was not discussed. Presumably there is some sort of tunnel on the ERV for transfer of the crew to the ERV? The details of the rendezvous with the ERV were not discussed at all. Hopefully this will be covered in a future publication by this group.

The mission plan assumed that the combination of the ERV and the landed payload would be placed into a "250 km $\times$ 33,800 km Mars orbit, commonly referred to as a 1 sol orbit" on arrival at Mars. The landed system would separate and descend to the surface, leaving the ERV in this elongated elliptical orbit.

Ascent from Mars take place through a series of steps as indicated in Table 1.2. P15 also presented a diagram of the steps, as shown here in Fig. 1.2. A close-up of the orbit raising steps is given in Fig. 1.3.

The whole process of ascent and rendezvous involves two major burns and five minor burns. The initial ascent and insertion into low circular orbit takes about a minute and requires a Δv of about 4 km/s. The main engines of the first stage provide for steps 1–4, and continue into the beginning of step 5, but the transition from stage 1 to stage 2 takes place at 3 min and 45 s after launch, prior to step 4. Stage 2 provides propulsion for steps 4 through 7.

Propulsion for the ascent vehicle was described as follows:

> The MAV propulsion system is separated into two stages. The first stage contains three 100 kN engines that consume liquid oxygen (oxidizer) and liquid methane (fuel). The engine operates at a specific impulse of 360 s and a propellant oxidizer-to-fuel mixture ratio of 3.5. The second stage contains a single main engine (same engine as first stage) along with four banks of four (16 total) reaction control engines. Each RCS engine produces 445 N of thrust and a specific impulse of 340 s. The corresponding mixture ratio is 3.0. The first stage MPS pulls propellant from two nested tanks on that stage. The second stage tanks provide propellants to both the MPS and RCS engines. RCS propellants must be vaporized and then pumped into accumulator tanks before usage. Gaseous propellants allow high pulse rates during attitude control. The maximum pressure in the accumulator tanks is 27.6 MPa (2760 bar). The gas-fed engine assumption proved to be complex and resulted in a high power requirement. Therefore, liquid-fed RCS engines will be studied for future design iterations.

Table 1.2 Steps in ascent and rendezvous

	Propulsive step	Periapsis height (km)	Apoapis height (km)	Δv (km/s)	Time from launch (h: min)	Color of orbit in diagram
1a	Launch			4040[a]	00:00	Black
1b	Circularize to 166 km orbit	166	166			Blue
2	Raise periapsis	250	166	15	00:57	Blk-dash
3	Raise periapsis	471	162	50	01:51	Green
4	Adjust phasing to enable rendezvous	470	200	8	04:43	Orange
5	Co-elliptic orbit	33,792	200	1147	24:50	Red
6	Target for rendezvous	33,792	240	4.6	34:50	Black
7	Braking for final rendezvous	33,796	250	1.7	43:55	Brown
	Total			5270		

[a]The Δv for this first step was not given directly in P15 but can be calculated to be 4.04 km/s since the total Δv was given as 5.27 km/s, and the sum of transfers after the initial lift to Mars orbit requires 1.23 km/s)

The first stage includes two tanks of methane holding 2298 kg each, and two tanks of oxygen holding 7841 kg each. Total methane is 4.6 mT and total oxygen is 15.7 mT. Total propellants are 20.3 mT.

The second stage includes two tanks of methane holding 1191 kg each, and two tanks of oxygen holding 3523 kg each. Total methane is 2.4 mT and total oxygen is 7.0 mT. Total propellants are 9.4 mT.

For a crew of four, estimated requirements were total methane: 7.0 mT, total oxygen: 22.7 mT and total propellants: 29.7 mT.

It should be noted that the ratio of dry mass to propellant mass for stage 1 was 15.8%. The ratio for the second stage was 25.5%.

The thrusters are pressure-fed from propellants stored at 50 psi, and the estimated specific impulse was 360 s. The tank pressures are maintained at 345 kPa (34.5 bar). The LOX is pressurized with helium, stored in ambient bottles at 31 MPa (3100 bar).

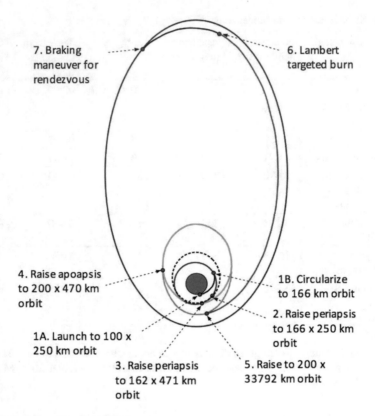

7. Braking
maneuver for
rendezvous

6. Lambert
targeted burn

4. Raise apoapsis
to 200 x 470 km
orbit

1B. Circularize
to 166 km orbit

2. Raise periapsis
to 166 x 250 km
orbit

1A. Launch to 100 x
250 km orbit

5. Raise to 200 x
33792 km orbit

3. Raise periapsis
to 162 x 471 km
orbit

Fig. 1.2 Propulsive steps for rendezvous

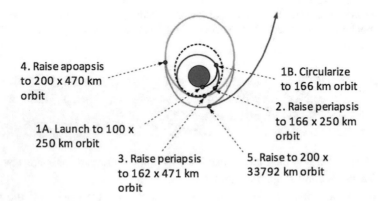

4. Raise apoapsis
to 200 x 470 km
orbit

1B. Circularize
to 166 km orbit

2. Raise periapsis
to 166 x 250 km
orbit

1A. Launch to 100 x
250 km orbit

5. Raise to 200 x
33792 km orbit

3. Raise periapsis
to 162 x 471 km
orbit

Fig. 1.3 Propulsive steps orbit raising

Cryocoolers maintain the propellant temperatures to prevent boil off. The liquid oxygen is kept at 94 K, while the liquid methane is stored at 112 K. Other assumptions include a 5% ullage volume for each propellant, 1% residual propellants, and a 1% fuel bias. P15 provided detailed estimates of the masses of the propulsion system components.

The required mass of ascent propellants is roughly proportional to the mass of the ascent capsule. It seems likely that the required mass of ascent propellants for a crew of six will be roughly 45 mT.

As I have discussed previously, a critical factor in determining the design and requirements for the MAV is the crew size. While several previous mission designs called for a crew of six, the current NASA mode envisages a crew of four in order to make the mission more affordable, although this introduces a number of risks and constraints on what can be accomplished by the crew during the mission. In keeping with current NASA policy, P15 assumed a landed crew of four. They provided 17.5 m^3 of internal pressurized volume for the MAV (equivalent to a $\sim$1.6 m radius sphere).

The dry mass of the MAV was given in Table 1.1 of P15. It is shown here as Table 1.3.

Further details are given in Table 1.4.

The masses in these tables have imbedded in them, allowance for future growth of $\sim$25% over current estimates. According to JPL design principles, 25% is inadequate at this early stage of design.

A number of aspects of the system were not discussed. These include the following:

(1) Presumably, the MAV sits atop a platform above the descent system, which forms a base for the MAV? This layout was used in some previous studies. If that were so, the layout of the MAV and its tunnel, and its connection to the cryocooling system and power system would be of interest.

(2) While P15 was mainly concerned with the MAV itself and not the source of propellants, nevertheless the source of propellants (whether generated all or partly by ISRU, or brought from Earth) is relevant. P15 was not very clear on this point. Obviously, if some, or all-cryogenic propellants are to be brought from Earth, maintaining them without excessive boil-off presents a challenge during descent and the set up time on the surface. A rather enigmatic paragraph in P15 is difficult to comprehend:

Shortly after arrival into a 250 km × 33,800 km Mars orbit, commonly referred to as a 1 sol orbit, the lander carrying the MAV detaches from the Earth-Mars transportation system. After a brief period of final checkouts and phasing to align with the targeted landing site, the lander descends to the surface. Once on the surface the lander must be connected to a surface power generator, currently assumed to be a fission power source that is delivered with the MAV or on an earlier lander. Connection to surface power is assumed to occur within 24 h after landing. Surface power is required for liquid oxygen production and to prevent boiloff of the ascent propellants. Oxygen production rates are dependent on available power, but the

Table 1.3 Dry mass of MAV according to P15

Subsystem	Mass (kg)
Crew cabin	3846
Structures	858
Power	377
Avionics	407
Thermal	542
ECLS	298
Cargo	1106
Non-prop. fluids	258
1st stage dry mass	3195
2nd stage dry mass	2378
Total dry mass	13,265

Table 1.4 Dry mass of MAV according to details in P15

Subsystem	Mass (kg)
Capsule structure	858
Tank and thruster supports	355
1st stage dry mass	2933
2nd stage dry mass	1559
RCS system	676
Power	377
Fuel cell reactants	333
Thermal	542
Coolants	63
Avionics	407
ECLS	298
Fluids for ECLS	196
Cargo	1106
Total capsule	9703

production duration is expected to be up to one year. During surface operations the MAV relies on the descent stage for communications, thermal control, and connection to surface power.

In the "split mission" approach, the MAV and power system are delivered at a previous launch opportunity. If all the ascent propellants ($CH_4 + O_2$) are generated using ISRU, ISRU processing will not begin until about a month or so after landing, needed to deploy, set up, and connect the power system. But if ISRU produces only O_2, the MAV lands with cold CH_4 in its tanks, and there is an immediate need for cryocooling these tanks. But that would have occurred more than a year prior to lift-off from Mars. A smaller power system for cryocooling that could be set up rapidly would likely be needed.

It is instructive to calculate some gear ratios. The ratio of total MAV mass at liftoff to the capsule mass delivered to the 1 sol orbit is 39.1/3.8 = 10.3. The ratio of total MAV mass at liftoff to the mass transferred from the MAV into the ERV (crew, suits, samples $\sim$700 kg) is 39.1/0.7 = 55.9.

The (propellant mass/dry mass) for a crew of 4 was estimated to be 29.7/13.3 = 2.23. For a crew of six, I simply multiplied the dry mass for a crew of four by 1.5. The propellant mass is assumed to be proportional to the total dry mass. The dry mass for a crew of four was 13.3 mT and the propellant mass was 29.7 mT. For a crew of six I then roughly estimated the dry mass to be 20.0 mT and the propellant mass is 44.6 mT, of which about 33.5 mT is Oxygen.

> For a crew of six, total methane is 11.1 mT, total oxygen is 33.5 mT and total propellants are 44.6 mT.

In a very recent paper, Polsgrove et al. (2017) reached, more or less, the same conclusions that we reached here:

> The total ascent vehicle mass drives performance requirements for the Mars descent systems and the Earth to Mars transportation elements. Minimizing Mars Ascent Vehicle (MAV) mass is a priority and minimizing the crew cabin size and mass is one way to do that.
> For the MAV, the crew cabin size and mass can have a large impact on vehicle design and performance.

In the 2017 paper, they explored a greater range of ascent propellants, staging and rendezvous orbits. Some of the numbers changed slightly from the 2015 paper for ascent propellants based on a crew of four (oxygen/methane now 25.0/7.9 vs. 22.7/7.0 in 2015). Various physical configurations for the ascent vehicle were examined. We have not reviewed this work in detail because from the point of view of ISRU, the main issue is simply required mass of ascent propellants. This work by Tara Polsgrove and co-workers over the past few years represents the first significant progress in designing and analyzing Mars Ascent Vehicles. Without their work, we would be mainly guessing as to the propellant requirements for ascent, and therefore guessing about the role of ISRU in a human mission to Mars.

1.1.5   Required ISRU Production Rates for Ascent Propellants

The cargo mission to deploy the power system, ascent vehicle and ISRU system to the surface of Mars departs at a certain date we can call "X". It is likely to be sent on a low energy trajectory requiring about 9 months. Thus, cargo landing takes place at (X + 9) months. The nuclear power system needs to be deployed at a great distance from the MAV, and a cable will connect the two. The cryogenic system must cool down the MAV tanks. All told, an optimistic estimate is that it would take at least a month to get the ISRU powered up and working around the clock. Thus, optimistically, ISRU is initiated at (X + 10) months. The ISRU system fills the MAV tanks during the following 14 months.

The next opportunity for launch to Mars occurs at (X + 26) months. This opportunity would be used to send the crew to Mars if the conditions were judged to be appropriate: A fully fueled MAV was waiting on the surface, and a competent ERV was waiting in Mars orbit. Once the decision was made to send the crew at this juncture, about 2–3 months of final preparation would be needed prior to launch. Therefore, the MAV tanks would have to be full at around (X + 23) to (X + 24) months. Thus the time available to fill the tanks of the MAV would be 13–14 months. Taking the optimistic view, we assume here that the ISRU system has 14 months of operation available to fill the MAV tanks.

As we discussed in previous sections, the propellant requirements for ascent have been estimated, and despite the excellent study by Polsgrove et al. (2015), it is still far too early to nail down these requirements accurately. The crew size has a major impact on ascent propellant requirements. The current best estimates are given in Table 1.5 for a full scale ISRU system. If methane is brought from Earth, only the oxygen needs to be produced by ISRU.

For ISRU systems that produce only oxygen from $CO_2 \Rightarrow CO + O$, the rate at which CO_2 must be provided depends on the following:

(i) Two CO_2 molecules are required to generate one O_2 molecule. Thus 88 g of CO_2 are needed to generate 32 g of O_2 at 100% conversion.

(ii) The percentage utilization of CO_2 may vary from about 50% for solid oxide electrolysis to about 95% for the Sabatier process.

Therefore, for solid oxide electrolysis production of oxygen, the required CO_2 delivery rate for a crew of four is about 2.3 kg/h × (88/32) × (1/0.5) = 12.7 kg/h. For a crew of six it is about 19 kg/h. If some of the pressurized CO_2 in the electrolysis tail gas can be recovered and recirculated, this would be reduced, possibly significantly. For a process that utilizes 95% of the CO_2, the required CO_2 delivery rate for a crew of four is about 2.3 kg/h × (88/32) × (1/0.95) = 6.7 kg/h. the required CO_2 delivery rate for a crew of six it is about 10 kg/h.

Table 1.5 Requirements for a full scale ISRU system

Crew size	Oxygen required (mT)	Methane required (mT)	Oxygen production rate (kg/h)	Methane production rate (kg/h)
4	22.7	7.0	2.3	0.7
6	33.5	11.1	3.3	1.1

1.1.6   Life Support and Consumables

1.1.6.1   Consumable Requirements (Without Recycling)

Life support during the three major legs of a Mars human mission (transit to Mars, surface stay, and return to Earth—and also descent and ascent) poses major challenges for human missions to Mars. The estimated total consumption of consumables for a crew of six for a round trip to the surface of Mars exceeds 100 metric tons (mT) and might be as much as ~ 200 mT. This could require an IMLEO of over 2000 mT, or roughly 13 launches with a heavy-lift launch vehicle just for life support consumables if neither recycling nor use of indigenous water from Mars were used. Clearly, life support is a major mass driver for human missions to Mars, and recycling and possibly use of indigenous Mars water resources are elements of any rational plan to make such missions feasible and affordable.

Life support, as defined by the NASA Advanced Life Support Project (ALS), includes the following elements:

- Air supply
- Biomass production
- Food supply
- Waste disposal
- Water supply.

Conceptually, each of these elements participates in a comprehensive overall environmental control and life support system (ECLSS) that maximizes recycling of waste products. These systems are complex and highly interactive.

Consumption requirements are summarized in Table 1.6. These estimates were derived from NASA Advanced Life Support (ALS) reports but further refinement is needed, particularly for water. For a crew of six over a complete mission, the total requirement is $6 \times 990 \times 33.6 = 200{,}000$ kg $= 200$ mT.

In order to characterize ECLSS for a human mission to Mars, a first step would be to catalog the inventory of consumables that are needed for each leg of the trip to support a crew of six, assuming no ECLSS is utilized. One would tabulate how much food, water (various qualities), oxygen, atmospheric buffer gas and waste disposal materials are needed, first on a per-crew-member-per-day basis, and then for the whole stay for a crew of say, 6. Unfortunately, this basic information is not presented in any of the ALS reports. Therefore, I have estimated this data as shown in Table 1.6. The resultant gross life support consumptions for a human mission to Mars are summarized in Table 1.7 assuming no use is made of recycling or indigenous resources. The total mass consumed over all mission phases is 200 mT.

1.1.6.2   Use of Recycling Systems

The NASA Advanced Life Support (ALS) Program is actively working on processes and prototype hardware for Environmental Control and Life Support Systems (ECLSS) that

Table 1.6 Estimated consumption requirements for long-term missions

Item	Requirements [kg/(person-day)]
Oral hygiene water	0.37
Hand/face wash water	4.1
Urinal flush water	0.5
Laundry water	12.5
Shower water	2.7
Dishwashing water	5.4
Drinking water	2.0
Total water	27.6
Oxygen	1.0
Buffer gas (N$_2$?)	3.0
Food	1.5
Waste disposal mtls	0.5
Total consumption	33.6

Table 1.7 Gross life support requirements for a human mission to Mars without recycling or in situ resource utilization for a crew of six (Metric tons)

Mission phase	Transit to Mars	Descent	Surface stay	Ascent	Earth return
Duration (days)	180	15	600	15	180
Water	29	2	100	2	29
Oxygen	1.1	0.3	4	0.3	1.1
Food	1.6	0.1	5.4	0.1	1.6
Waste disposal Mtls	0.6		1.8		0.6
Buffer gas	3.3	0.9	12	0.9	3.3
Total consumed	36	3	123	3	36

involves recycling vital life support consumables so as to reduce the mass that must be brought from Earth for human space missions.

Not only must the ECLSS provide the gross requirements for these elements, but it must also monitor trace contaminants and reduce them to an acceptable level. NASA life support data are reported in two segments. One segment is claimed to be "state of art" based on "the International Space Station (ISS) Upgrade Mission" and the other segment is for an Advanced Life Support system (ALS) that is based on advanced technologies currently under development within NASA. It is claimed that technologies included in the assessments have (as a minimum) been taken to the breadboard test stage. The ALS reports provide numerical estimates for mass and power requirements of ECLSS systems. However, the connection between the baseline data in the reports and actual experience with the ISS is difficult to discern. It is not clear how much experimental data underlie the tables, and how much data are estimated from modeling. Nor is it clear whether these systems are reliable for the long transits and surface stays of Mars missions (most of this work seems to address systems in Earth orbit). The longevity and mean time between failures of these systems has not yet been reported. It is noteworthy that all of the mass estimates provided by the ALS are from the research arm of NASA, and they do not include allowances for margins, redundancy or spares.

Most of the available reports provide system estimates of masses for the various elements of the life support system (LSS). The basic element masses are listed, as well as "equivalent system masses" (ESM) that include additional mass to account for the required power systems, thermal systems and human oversight requirements associated with operation of the LSS. However, we do not use equivalent system masses herein. Therefore, our mass estimates are lower bounds.

The major elements that are recycled are breathing air and water. For each recycle system, the mass of the physical plant needed to supply the consumables must be estimated, as well as the recovery percentages for the air and water systems. From the recovery percentage, one can calculate the size of the back-up cache needed for replenishment of lost resources during recycling. Then, for each of the air and water systems, five quantities would be reported for each mission leg:

1. The total mass of the resource needed for a crew of six over the duration of the mission leg (M_T)
2. The mass of the physical plant (M_{PP})
3. The recovery percentage (R_P) (percent of used resource that is recovered in each cycle)
4. The mass of the back-up cache needed for replenishment of losses in recycling: $M_B = (100 - R_P) M_T/100$
5. Total mass of the ECLSS that supplies M_T of resource during the mission leg: (sum of physical plant + backup cache): $M_{LS} = M_{PP} + M_B$.

A useful figure of merit is the ratio M_T/M_{LS} that specifies the ratio of the mass of resource supplied to the total mass of the ECLSS system. The larger this ratio, the more efficient is the ECLSS.

In addition to these performance estimates, the reliability and longevity of such systems should be discussed, and additional mass provided for margins, spares, and redundancy, as needed.

Finally, the potential impact of utilizing indigenous water on Mars for surface systems should be considered and incorporated into plans as appropriate.

Only water and air are susceptible to recycling, whereas the elements biomass production, food supply, waste disposal and thermal control are not recycled.

We hypothesize a mission with a crew of six, 180-day transits to and from Mars, and a 600-day stay on Mars. The requirements were estimated in Table 1.7 assuming no recycling. Using NASA estimates for recycling, we obtain Table 1.8. The total mass delivered for each mission phase is the sum of ECLSS masses for air and water, plus food and waste disposal materials, although no recycling is assumed for the short duration ascent and descent steps.

Table 1.8 Estimated requirements for life support consumables using NASA estimates for ECLSS for a crew of six, without use of ISRU. Masses are in mT

Mission phase	Transit to Mars	Descent	Surface stay	Ascent	Earth return
Duration (days)	180	15	600	15	180
Water requirement	29	2	100	2	29
Water ECLSS plant mass	1.4		4.1		1.4
Water ECLSS recovery %	>99		94		>99
Water ECLSS back-up cache mass	0.3		6.3		0.3
TOTAL water ECLSS mass	1.7	2	10.4	2	1.7
Water mass/ECLSS mass ratio	17	1	10	1	17
Air requirement	4	0.9	12	0.9	4
Air ECLSS plant mass	0.5		1.3		0.5
Air ECLSS recovery %	83		76		83
Air ECLSS back-up cache mass	0.7		2.9		0.7
TOTAL air ECLSS mass	1.2	0.9	4.2	0.9	1.2
Air mass/ECLSS mass ratio	3	1	3	1	3
Food	1.6	0.15	5.4	0.15	1.6
Waste disposal materials	0.5	0.05	1.8	0.05	0.5
Total mass delivered to Mars	5.0	3.1	21.8	3.1	5.0

This might add to the required back-up cache and/or require some spares or redundant units that would double (or more) the mass of the system. Obviously, long-term testing is needed here. Sanders and Duke (2005) emphasized the need for ISRU as a backup for an ECLSS, pointing out the potential unreliability of ECLSS. It is also interesting that the NASA ALS appears to be rather cautious regarding the potential for widespread indigenous Mars water resources to impact life support on Mars, despite the fact that this impact could potentially be a major benefit in mass reduction and safety. Admittedly, acquisition of such water resources will require a significant investment and there are concerns regarding planetary protection. Nevertheless, this aspect would seem to deserve more attention in ALS activities.

A NASA report claimed:[2]

> Experience with Mir, International Space Station (ISS), and Shuttle, have shown that even with extensive ground checkout, hardware failures occur. For long duration missions, such as Mir and ISS, orbital replacement units (ORUs) must be stored on-orbit or delivered from Earth to maintain operations, even with systems that were initially two-fault tolerant. Long surface stays on the Moon and Mars will require a different method of failure recovery than ORUs.

---

[2]http://www.marsjournal.org/contents/2006/0005/files/SandersDuke2005.pdf.

Use of indigenous water on Mars may provide significant mass savings as well as great risk reduction. It is conceivable that the entire water supply needed while the crew is on the surface could be supplied from near-surface indigenous resources, thus eliminating water recycling altogether on the surface of Mars. In addition, the entire surface oxygen supply could be provided as well. The only commodity needing recycling on Mars would then be atmospheric buffer gas.

It is hoped that in the future, the ALS will:

(1) Concentrate on systems with very high reliability for long durations rather than systems with very high recovery percentages. For Mars, a LSS with 90% recovery and 99.8% reliability would be far more valuable than one with 99.8% recovery and 90% reliability.
(2) Provide clearer delineation of data sources with particular emphasis on which data are based on experiment, and what the duration of the experiments were.
(3) Consider use of the widespread near-surface water resources on Mars.

1.1.6.3   Life Support Summary

In summary, we can draw the following conclusions regarding life support consumables for Mars missions:

- Although there are half a dozen elements involved in life support, water is by far the greatest factor in determining the mass requirements for life support.
- If no recycling or use of indigenous resources is used, the requirements for consumables in a human mission to Mars would be about 200 mT, which in turn, might require over 2000 mT for IMLEO. Such a scenario would be prohibitive and is totally impractical. About half of the 200 mT would be needed during the stay on the Mars surface.
- If recycling is used for air and water, and ALS mass estimates for ECLSS based on ISS experience are adopted, the total mass brought from Earth decreases from ~ 200 to 38 mT, and IMLEO decreases from over 2000 mT to perhaps 400 mT. This would still require several heavy-lift launches solely for life support consumables and recycling plants.
- The connection between ALS estimates of ECLSS system masses and actual ISS experience has not been divulged. Therefore the experimental basis for ALS estimates is unclear.
- The ALS reports do not discuss longevity, reliability and mean time between failures of ECLSS. It is not clear how transferable ISS data are to Mars missions where ECLSS must function without failure for 2.7 years.
- The estimates provided herein do not include allowances for margins, redundancy or spares.
- Use of indigenous Mars water resources has the potential to eliminate or significantly simplify the need for recycling of water and oxygen on the surface of Mars.

1.1.7   Mars Surface Transportation

A critical part of any human mission to Mars will involve fairly long-distance travel by some of the crew to explore Mars surface features. Most DRMs did not elaborate on the designs and requirements for surface transportation. How many crewmembers? What range of distance? What cargo and capabilities? Will there be pressurized and unpressurized rovers? How will power, life support and navigation be implemented?

Perez-Davis and Faymon (1987) reported on a study of a manned Mars rover that used hydrogen-oxygen fuel cells as the primary power system. The vehicle performance requirements included 10 km/h speed, 100 km range, 30° slope climb for 50 km, mission duration 5 days, and crew = 5. Vehicle masses ranging from 2700 to 7600 kg were examined. The fuel cell system was derived from the Shuttle hydrogen-oxygen alkaline fuel cell designs. Water produced by the fuel cell is stored for return to the base. Heat generated by the fuel cell is used for thermal control of the interior. While several tables of data were provided, the overall design and operation seem sketchy.

Rowland et al. (2004) presented results of a study by a student research team. They briefly described the unpressurized lunar rovers built by Boeing. For Apollo 17 the rover weighed 209 kg and could carry up to 517 kg of payload. This was a one-time use rover and was not designed for long-term use. Non-rechargeable silver zinc batteries provided power (115 A-h). For Mars they considered pressurized rovers that could go well beyond walking range of crew. Thus at least two rovers would be required for rescue in case of a mishap. They estimated that a pressurized Mars rover capable of carrying three crewmembers for a 14-day mission and weighing 5000 kg will require 2000 kW-h of stored energy. The power system will experience a peak load of 90 kW when the rover is running life-support, computers, communications, and driving uphill over rocks. They examined possible energy sources: nuclear fission, solar, batteries, internal combustion engines, and fuel cells. Fuel cells were the obvious choice but questions were raised about durability. Alkaline fuel cells were chosen over PEM and solid oxide versions. Heat removal from the rover was identified as a potential challenge. The airlock design poses major challenges as well. Cohen (2000) analyzed various approaches for airlock design.

Bradshaw (2008) presented a conceptual design of a small pressurized rover to carry two crewmembers, capable of multi-day or week-long sortie durations, for use on the Moon, but possibly extensible to Mars. This design featured:

- Suitports to minimize gas loss and dust contamination in suit donning and egress.
- 2.5 cm of ice as radiation protection.
- Modular tool attachment for various jobs.
- EVA driving station allows crewmember to drive while outside pressurized volume.

NASA has published a concept for a pressurized rover designated as the Space Exploration Vehicle.[3]

There is no doubt that a vital part of any human missions to Mars is providing long-range surface transportation for crewmembers. It seems almost certain that a pressurized rover will be required. Until now, designs of such machines have been only conceptual. Providing power for pressurized rovers is inherently tied to ISRU, but the details have yet to be worked out.

1.2   Mars Resources

The principal applications to use ISRU on Mars include propellants for ascent propulsion from the Martian surface, and water and oxygen for life support. Mars soil contains oxygen but processing the soil is energy intensive and mechanically challenging. The Mars atmosphere is ubiquitous and provides a source of carbon and oxygen in the form of $\sim 95\%$ CO_2. However, the average pressure is less than 1% of 1 atmosphere and therefore, the CO_2 must typically be compressed by about a factor of 30–100 for processing in reactors of practical size. In order to separate the carbon from the oxygen in CO_2, a reducing agent is required. Hydrogen is the obvious choice, and when hydrogen is reacted with CO_2, CH_4 and O_2 are produced. Unfortunately, hydrogen is a scarce commodity on Mars. While it might be technically feasible to transport hydrogen from Earth to Mars, this is a very difficult proposition and is probably impractical due to mass, volume and power requirements. Storing the hydrogen on Mars after it gets there is even more difficult. There are strong indications that near-surface H_2O might be available in various regions of Mars, and if this is verified, mining these H_2O deposits might be practical. In this case, a combination of mined near-surface H_2O plus compressed atmosphere might provide a viable source of carbon, hydrogen and oxygen for processing.

1.2.1   The Atmosphere

One fundamental resource on Mars is the atmosphere, containing roughly:

Carbon dioxide	95.32%
Nitrogen	2.7%
Argon	1.6%
Oxygen	0.13%
Carbon monoxide	0.07%
Water vapor	0.03%
Nitric oxide	0.013%

[3]https://www.nasa.gov/pdf/464826main_SEV_Concept_FactSheet.pdf, https://www.nasa.gov/exploration/technology/space_exploration_vehicle/index.html.

The pressure of the atmosphere depends on the local elevation and the season but a rough average value is about 0.007 atm.

Most schemes for implementing ISRU on Mars utilize the CO_2 in the atmosphere, but typically this CO_2 must be compressed about by a factor of at least 30 in order to reduce the size of chemical reactors to manageable size. The compressed CO_2 then acts as a feedstock for chemical processing. Some schemes also collect the inert gases N_2 and Ar for use as a diluent for oxygen in life support systems. A scheme has also been proposed to recover the small amount of H_2O in the atmosphere.

In utilizing the CO_2 in the Mars atmosphere, the intake system must be designed to reject prevailing dust that might create problems within processing systems.

1.2.2   Near-Surface H₂O

Appendix III of Rapp (2015) provides an extensive review of evidence for existence of near-surface H_2O in some regions of Mars. The H_2O at higher latitudes is believed to be ground ice whereas at lower latitudes it is most probably water of hydration of minerals in the regolith. In this context, "near-surface" implies within the top meter or two of regolith.

A conflux of theoretical models and experimental data provide a very strong indication that near-surface subsurface H_2O is widespread on Mars at higher latitudes, and experimental data indicate that some form of H_2O reaches down to near-equatorial latitudes in some regions.

Equilibrium Models:

We are directly aware of H_2O on Mars by observing the polar caps and by measuring concentrations of water vapor in the atmosphere. The water vapor interacts with the cold porous surface and may (depending on temperatures and water vapor concentrations) act as a source to deposit H_2O into the porous subsurface, or act a sink to withdraw H_2O from the subsurface. This process has been extensively modeled by a number of scientists.

At any location on Mars, if the asymptotic subsurface temperature is low enough that the vapor pressure (over ice) at that temperature is lower than the water vapor partial pressure of the atmosphere, water will tend to diffuse through the regolith and de-sublimate out as ice in the subsurface. This is believed to be a fairly slow process, but is rapid enough to allow significant transfers of ground ice to and from the atmosphere over obliquity cycles (tens of thousands of years).

Given that in some locations, the asymptotic subsurface temperature is low enough and the atmospheric partial pressure of water vapor is high enough that subsurface ice will form, the minimum depth at which subsurface ice is stable depends upon the temperature profile as the temperature changes from the surface temperature downward to the asymptotic temperature at deeper depths. A number of authors have made detailed esti-mates of the temperature profiles (that depend on thermal inertia of subsurface), and they have estimated minimum depths for stable ice formation ("ice table") for various soil properties and latitudes [see Rapp (2015) for specifics].

Although details vary from author to author, the general outlines of models are similar. These can be summarized as follows:

(a)   The mean annual temperature in the equatorial region ($-30°$ to $+30°$ latitude) is too warm and the water vapor concentrations in the equatorial atmosphere are too low to allow subsurface ground ice to be thermodynamically stable under present conditions. Subsurface ice in this region will gradually sublime away over time. However, it is possible (but very unlikely) that previously deposited ground ice from a past ice age might take a long time to disappear and might remain in a non-equilibrium state for some considerable time in a few unique locations—and possibly be present today.

(b)   At very high latitudes, the year-around temperatures are low enough and the average water vapor concentration is high enough to support stable ground ice at the surface and below the surface. The surface ice cap grows and retreats with the seasons. During local summer in the northern hemisphere, a good deal of ice is sublimed, raising the water vapor concentration in the northern atmosphere. A lesser effect occurs in the southern hemisphere.

(c)   As one moves away from the poles, a latitude is reached (perhaps in the range $50°$–$60°$ depending on soil properties, local temperatures, local water vapor concentrations, slopes, etc.) where the depth of the equilibrium subsurface ice table increases sharply as one moves equatorward.

(d)   If the regolith is as porous as suspected, and if the measurements of water vapor and temperature on Mars are correct, significant amounts of subsurface ice must form in the pores of the regolith at higher latitudes by the laws of physical chemistry, and the demarcation line where near surface ice is no longer stable varies with terrain, soil properties and local weather, but is probably in the range $50°$–$60°$ latitude. The regions where subsurface ice remains stable depend on unknown subsurface prop-erties as well as surface elevation, orientation and composition.

(e)   These results for current equilibrium subsurface ice stability are remarkably different from past epochs when the Mars orbit tilt was much greater, thus enhancing polar heating by the sun and reducing solar heating in temperate zones. During those periods, near-surface ice may have been stable over much of Mars, and considerable subsurface ice could have been deposited in temperate zones. In addition, the effect of periodic precession of the equinoxes would have moved water back and forth between the poles, probably depositing ground ice at intermediate latitudes along the way.

Past Ice Ages:

As Appendix II of Rapp (2015) shows, the orbit of Mars has undergone rather large variations during the past million years. The most important factor is the variation in obliquity, but variations in eccentricity and periodic precession of the equinoxes are also relevant. Such variations produce major changes in the distribution of solar energy input to Mars versus latitude, potentially resulting in redistribution of H_2O resources over this period.

Mellon and Jakosky (1995) found that moderate changes in the Martian obliquity can shift the geographic boundary of stable ground ice from the equator (global stability) up to about 70° latitude, and diffusion of water vapor is rapid enough to cause similarly dramatic changes in the presence of ground ice at these locations on time-scales of thousands of years or less. It was estimated that the ice content of the upper 1–2 m of the soil can vary widely due to exchange of atmospheric water at rates faster than the rate of change of Mars obliquity. They provided an analysis of the behavior of near-surface ground ice on Mars through many epochs of varying obliquity during the past ~ 1 million years. They pointed out that the Mars climate undergoes two major responses to changing obliquity: (1) temperature change due to redistribution of insolation versus latitude, and (2) increased summertime water sublimation from polar caps during higher obliquity thus increasing the atmospheric water abundance and affecting the rate and direction of diffusive transport of water vapor in exchange with regolith at various latitudes. It turns out that the increase in atmospheric water abundance is more important than the temperature changes in regard to deposition of ground ice at equatorial and mid-latitudes. A comprehensive thermal/diffusion model allowed mapping out ground ice formation and depletion as a function of depth, latitude and Mars orbital history. Their model describes regions and time periods where ice is stable as well as regions and time periods where ice is not stable but previously deposited ice remains residual because insufficient time has passed to allow it to sublime.

According to this model, the mean atmospheric water vapor abundance would be about 35 times greater at an obliquity of 32° than it is today at 25.2°. This raises the frost point from ~ 195 to ~ 218 K, allowing stable deposition of ice in the regolith at all latitudes. Because of the non-linear dependence of water vapor pressure on temperature, and the direct dependence of diffusion on the water vapor pressure at the surface, when Mars enters into a period of increasing obliquity, there is a relatively rapid spread of ground ice into lower latitudes, culminating in planet-wide stable ground ice at sufficiently high obliquities. Subsequently, as the obliquity diminishes with time, the near-surface ground ice is gradually depleted due to sublimation. Therefore, during the past million years, it was concluded that there were periods of widespread ice stability alternating with periods where ice is stable only at high latitudes. Because oscillations of the obliquity of Mars have been relatively small during the past 300,000 years, this period has been marked by unusual stability.

It is noteworthy that this study found that "ice accumulates more rapidly during high obliquity than can be lost during low obliquity." Therefore, their curves of depth to the ice table tend to have a characteristic sharp reduction during the early stages of high obliquity, with a longer "tail" extending out as the obliquity diminishes. This appears to be due to the low subsurface temperatures that reduce the vapor pressure and rate of diffusion during the warming period as the obliquity diminishes. This study provides a great amount of data and it is difficult to summarize all of their findings succinctly. They determined the regions of ground ice stability and the depths of the ice table as a function of obliquity.

Chamberlain and Boynton (2004) investigated conditions under which ground ice could be stable on Mars based on the past history of changing obliquity of the Mars orbit. They used a thermal model and a water vapor diffusion model. The thermal model determines the temperatures at different depths in the subsurface at different times of the Martian year. Subsurface temperatures are functions of latitude, albedo and thermal inertia. The thermal model can determine the depth at which subsurface ice becomes stable. Ice is stable if the top of the "ice table" has the same average vapor density as the average water vapor density in the atmosphere. H_2O is allowed to move by the vapor diffusion model. The temperatures from thermal model are used to partition water between 3 phases: vapor, adsorbed and ice. Vapor is the only mobile phase and the diffusion of vapor is buffered by adsorbed water. Vapor diffusion models can have ice-poor or ice-rich starting conditions. Vapor diffusion models are run for long periods to check the long-term evolution of depth to stable ice. As the ice distribution in the subsurface changes, the thermal properties of the subsurface also change. Thermal conductivity increases as ice fills the pore spaces. Vapor diffusion models are run iteratively with thermal models to update the temperature profiles as ice is re-distributed. In one set of results, Chamberlain and Boynton (2004) presented data on stability of ground ice versus latitude for various Mars obliquities. The obliquity of Mars has varied considerably in the past. Two sets of ground properties were utilized:

(a) bright, dusty ground: (albedo = 0.30 and thermal inertia = 100 S.I. units)
(b) dark, rocky ground (albedo = 0.18 and thermal inertia = 235 S.I. units).

Their results indicate that ground ice is never stable at equatorial latitudes at low obliquity. However, as the obliquity is increased, a point is reached (depending on the soil properties) where a discontinuous transition occurs from instability to stability of ground ice. According to this model, this transition occurs at obliquities between 25° and 27° for bright dusty ground, and between 29° and 31° for dark rocky ground. With the present obliquity at 25.2°, Mars is at the ragged edge of the realm where ground ice could be stable in some locations at equatorial latitudes. According to Appendix II of Rapp (2015), there were several periods in the past million years when the obliquity reached 35°. Even in the past ∼400,000 years, the obliquity reached 30°, and was as high as 27° only ∼80,000 years ago. During those periods, solar energy input to equatorial regions was significantly reduced in winter and solar energy input to high latitudes was significantly increased in summer. It seems likely that there must have been a major transfer of near-surface ice from the high latitudes to the temperate latitudes during these epochs. The obliquity has been ≤25.2° over the past ∼50,000 years, implying that the subsurface ice deposited in earlier epochs has been subliming, receding, and transferring to polar areas.

However, these processes might be slow, and could be severely inhibited by dust in some localities. Therefore it is possible that in some very bright, low thermal inertia regions in the equatorial belt, some of this vestigial subsurface ice from former epochs may remain even today, particularly on surfaces tilted toward the poles. This could possibly explain the areas of higher water content in some regions of the near-equatorial belt observed by Mars Odyssey.

Observations from Orbit:

The Mars Odyssey orbiter utilized a Gamma Ray Spectrometer to detect hydrogen to a depth of the upper meter of Mars. Details on the instrument and measurements are given by Rapp (2015). The hydrogen was interpreted as due to water in some form. They estimated the percent water content in the upper meter of Mars across most of the planet. The Odyssey Gamma Ray Spectrometer made measurements of H_2O content all over Mars to depths of ~ 1 m, in area elements $5° \times 5°$ (300 km $\times$ 300 km). These measurements support the models that predict near-surface subsurface ice will be prevalent and widespread at latitudes greater than about 55–60°. The data also indicate pockets of locally relatively high H_2O concentration (8–10%) in near-equatorial regions where albedo is high and thermal inertia is low, suggestive of the possibility that some remnant ice, slowly receding from previous ice ages, might still remain near the surface.

The equatorial regions with relatively high water content (8–10%) present an enigma. On the one hand, thermodynamic models predict that subsurface ice is unstable near the surface in the broad equatorial region. On the other hand, the Odyssey data are suggestive of subsurface ice. It is possible that this is metastable subsurface ice left over from a previous epoch with higher obliquity. Alternatively, it could be soil heavily endowed with salts containing water of crystallization. The fact that these areas coincide almost perfectly with regions of high albedo and low thermal inertia suggest that it may indeed be subsurface ice. Furthermore, the pixel size of Odyssey data was several hundred km, and the $\sim 8\%$ water average figure for a large pixel might be due to smaller local pockets of higher water concentration (where surface properties and slopes are supportive) scattered within an arid background. Over the past million years, the obliquity, eccentricity and precession of the equinoxes of Mars has caused a variable solar input to the planet in which the relative solar input to high and low latitudes has varied considerably. Certainly, ground ice was transferred from polar areas to temperate areas during some of these epochs. It is possible that some of this ground ice remains today even though it is thermodynamically unstable in temperate areas. In order for remnant subsurface ice from past epochs to be a proper explanation, the process of ice deposition must be faster than the process of ice sublimation in the temperate areas over time periods of tens to hundreds of thousands of years.

Jakosky et al. (2005) explored the quantitative connections between regolith water abundance and each of the physical properties that might be controlling the stability of ground ice left over from past ice ages: regions where higher peak or mean water vapor concentration occurs in the atmosphere, where there is lower surface temperature, where the location occurs on the northward side of the pole-facing slope, where there is low thermal inertia, and where there is high albedo. They found a notable lack of statistical correlation between water abundance and any of the parameters, either singly, or in groups. They concluded that there is no sufficiency condition; that no set of these parameters guarantees high water abundance. However, Rapp (2015) showed that a necessary condition for high water abundance is that equatorial regions of high water abundance are associated with higher values of peak water vapor concentration, lower surface temperatures, location below the tops of north pole-facing slopes, and generally somewhat lower values of thermal inertia and higher values of albedo. Only such regions have high equatorial water abundance. On the other hand, some equatorial regions that satisfy these conditions do not have high water abundance. Thus, these conditions are necessary but not sufficient.

Morphology of Craters:

The morphology of craters on Mars suggests that there may be huge ice reservoirs in the subsurface. If this proves to be correct, such reservoirs would act as a source for formation of subsurface ice at all depths above the source. Only a few models have been developed for such an occurrence. Stewart et al. (2004) found that many Martian craters have ejecta blankets that appear to have been formed by melted ice.

Rapp (2015) provided a review of the topic. Large Martian craters typically have an ejecta sheet, and some have a pronounced low ridge or escarpment at its outer edge. Craters of this type are referred to as "rampart" craters. It has been observed that in general, rampart craters account for a significant fraction of fresh craters on Mars. The majority of craters on Mars are degraded to the point of no longer displaying an ejecta morphology. But among those that do show an ejecta blanket, layered ejecta morphologies, including those which show ramparts, dominate. In any local area, it has been found that rampart craters do not form at crater diameters below a critical onset diameter, D_o, where this onset diameter varies with location. Thus, for example, in an area where D_o happens to be say, 4 km, all craters of diameter <4 km in that area will lack the marginal outflow and have the appearance of lunar craters. Craters with diameters >4 km in that area will mainly be rampart craters. It is widely believed that rampart craters are produced by impacts into ice-laden or water-laden regoliths, although an alternative explanation for the morphology of rampart craters is based on the interaction of dry ejecta with the atmosphere. Presumably, craters much reach a sufficient depth to hit local ground ice and thereby display the rampart structure.

The cornerstone of understanding of Mars rampart craters and the relationship of crater morphologies to subsurface H_2O, was laid down by Barlow and Bradley (1990) with a broad study of Martian craters over a wide range of latitudes and morphologies. Previous studies had led to inconsistent results, probably due to limited areal extent and limited photographic resolution. Barlow and Bradley undertook "a new study of Martian ejecta and interior morphology variations using Viking images from across the entire Martian surface in an attempt to resolve some of the outstanding controversies regarding how and where these features form." Seven different types of crater morphology were defined. Their results suggested that there is an upper desiccated layer, below which is an ice layer, and below that a brine layer in some latitudes, with a volatile-poor layer beneath all these. Very small impact craters never penetrate to the ice layer and produce primarily pancake morphologies. Small craters of sufficient size to reach the ice layer produce primarily single layer ejecta morphologies. Larger craters may excavate both ice and brine layers, producing multiple layer ejecta morphologies. Very large craters primarily excavate the desiccated region resulting in radial morphologies. Barlow and co-workers upgraded the original work of Barlow and Bradley in a series of later studies.

Observations After Mars Odyssey:

In a NASA news release, it was reported that the High Resolution Imaging Science Experiment camera on NASA's Mars Reconnaissance Orbiter took images of a fresh craters on Mars in late 2008 and early 2009. These images revealed frozen water hiding just below the surface of mid-latitude Mars by observing bright ice exposed at five Martian sites with new craters that range in depth from approximately half a meter to 2.5 m. The craters did not exist in earlier images of the same sites. Some of the craters show a thin layer of bright ice atop darker underlying material. The bright patches darkened in the weeks following initial observations, as the freshly exposed ice vaporized into the thin Martian atmosphere. One of the new craters had a bright patch of material large enough for one of the orbiter's instruments to confirm it is water ice.

The finds indicate water ice occurs beneath Mars' surface between 43 and 47°N, probably as a relic from a past ice age.

Unknowns Remaining:

The unknowns remaining in regard to near-surface H_2O on Mars include the following:

- How widespread are near-surface ice deposits on Mars such as have been observed via new impact craters?
- Is the observation of 8–10% H_2O over near-equatorial 300 km × 300 km pixels valid?
- How is this H_2O distributed within these pixels?
- Is this H_2O ground ice, mineral hydration or both?

1.3   Acquiring Compressed CO_2

The first step in almost all concepts for Mars ISRU is acquisition and compression of CO_2 from the Mars atmosphere. This system will inevitably divide into two subsystems: dust control and compression. The system must be rugged and reliable enough to operate for about 500–600 days on the surface of Mars under severe daily and seasonal temperature variations as well as continuing exposure to suspended dust and periodic episodes of dust storms and dust devils.

At first glance, compression of Martian atmosphere might be considered to be a simple process. As it turns out, that is far from the case. Three possibilities that immediately suggest themselves are a sorption process, a cryogenic process, and a mechanical compressor. None of these provide an immediate simple, reliable solution.

1.3.1   Compressors

1.3.1.1   Sorption Compressor

A sorption compressor contains virtually no moving parts and achieves its compression by alternately cooling and heating a sorbent material that absorbs low-pressure gas at low temperatures and drives off high-pressure gas at higher temperatures. By exposing the sorption compressor to the cold night environment of Mars (roughly 6 Torr and 200 K at moderate latitudes), CO_2 is preferentially adsorbed from the Martian atmosphere by the sorbent material while a good part of the 4.5% of other gases in the atmosphere is vented. During the day, when solar electrical power is available, the adsorbent is heated in a closed volume, thereby releasing relatively pure CO_2 at significantly higher pressures for use in a CO_2 conversion reactor. Such a sorption compressor was proposed in the early papers on ISRU [e.g. Ash et al. (1978) and Stancati et al. (1979)]. A breadboard version of a sorption compressor was demonstrated by Zubrin et al. (1995), although it did not resemble a realistic compressor because it had no vacuum jacket, no heat switch, and required an external roughing pump to remove unadsorbed Ar + N$_2$.

Presumably, a sorption compressor would absorb at night and release high-pressure gas during the day. For a continuous gas supply, gas storage would be needed in a pressure vessel.

A representative idealized operating cycle for a Mars Atmosphere Adsorption Compressor (MAAC) based on adsorption data from the literature is shown in Fig. 1.4, assuming an arbitrary working pressure of 815 Torr for the chemical reactor that is associated with the MAAC (Rapp et al. 1997). The compression ratio is 136:1 and the temperature swing is from 200 to 450 K. The net release of gas from the adsorbent is estimated at roughly 0.11 g CO_2 per g Zeolite. These data were taken by Rapp et al. (1997). However, data taken by other investigators (Finn et al 1996; Clark 1998) suggests that the holding capacity of 13X Zeolite might be considerably higher. Data taken at Lockheed Martin in the late 1990s suggests a capacity of 0.16 g CO_2 per g of Zeolite.

The four basic steps of a simple cycle are the following:

- (A → B) Heating at Constant Volume. In the early morning at Point "A" ($\sim$200 K, 6 Torr) with the Zeolite material fully saturated with CO_2, the sorption bed is heated to drive off CO_2 until eventually, at point "B" (340 K, 815 Torr), the pressure has risen to a point where processing of CO_2 is appropriate.

Fig. 1.4 Mars sorption
compressor operating cycle
between 6 Torr and 815 Torr
for 13X Zeolite

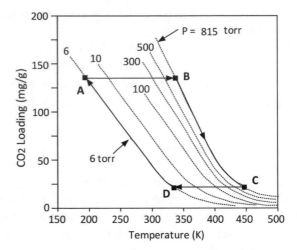

- (B → C) Heating at Constant Pressure. In the late morning, at point "B" (340 K, 815 Torr), a valve is opened allowing CO_2 to flow to the reactor. As the day progresses, the sorbent bed gradually becomes depleted, and it takes higher and higher bed temperatures to maintain the pressure at 815 Torr. By late afternoon, the sorbent bed is essentially depleted for all practical purposes (Point "C") and the heaters are turned off.
- (C → D) Cooling at Constant Volume. At the end of the day, at point "C" (450 K, 815 Torr) all valves are closed, and the cooling cycle begins. Heat from the bed is dumped to a radiator, cooling the sorbent material. Once point "D" is reached (340 K, 6 Torr), the pressure of the sorption compressor is equal to the ambient pressure and a valve can be opened to the Mars environment allowing adsorption to occur overnight.
- (D → A) Cooling at Constant Pressure. For the remainder of the evening, the sorbent bed cools as heat is dumped to the radiator. By the end of the evening, the sorbent material is fully saturated with CO_2 at 6 Torr and 200 K (point "A"), and the entire cycle is ready to begin again.

In order to provide a continuous supply of pressurized CO_2, several of these units would be operated in a more complex cycle.

The challenges involved in the design of a Mars sorption compressor include:

(1) Sorbent Material Characteristics—What quantity of CO_2 can be adsorbed and desorbed by a given amount of sorbent material, and what is the best material to use? Although various measurements have been made on adsorption by very small amounts of various kinds of sorbents, measurements of storage capacity in a larger scale configuration will determine which sorbent is most advantageous. There are four major concerns:

1. How much CO$_2$ the sorbent can adsorb at 6 Torr and ~200 K? [This determines the ultimate capacity of the sorbent.]
2. What fraction of the adsorbed gas can be released at ~750 Torr as a function of temperature above ~400 K? [This has a major effect on the power required to release the CO$_2$ from the sorbent.]
3. What is the behavior of the sorbent when exposed to a Mars gas mixture (95% CO$_2$, with the remainder made up mainly of Ar and N$_2$, and smaller amounts of CO and O$_2$)? [This is a very vital and relatively unstudied aspect of sorbents. It is clear that the trace elements can seriously interfere with adsorption of CO$_2$.]
4. Is Mars dust detrimental to the sorption material? If some dust is ingested into the sorption bed, will this affect its performance? [This determines to what extent filters must be employed on the gas intake.]

Two commercial Zeolites are known to adsorb reasonably large quantities of CO$_2$ at ~200 K and ~6 Torr (estimated to be in the range 15–18% by weight). These are the so-called 13X and 5A varieties.

(2) Removal of Non-CO$_2$ Gases—A system must be provided that prevents unadsorbed permanent gases from building up significant concentrations around the adsorbent material (or possibly in the pores of the material) as the CO$_2$ sorption process proceeds, thus creating a diffusive barrier to further adsorption of CO$_2$. Previous studies have shown experimentally that permanent gases such as Ar and N$_2$ are preferentially desorbed compared to CO$_2$ when the sorption bed is first heated. If this gas is vented, the remaining gas desorbed from the sorbent by heating can approach almost pure CO$_2$ as a feedstock for conversion to propellants. However, it has been observed that the less-adsorbed gases from the Martian atmosphere such as N$_2$ and Ar can inhibit the adsorption of CO$_2$. One theory is that these gases tend to build up significant concentrations around the sorbent material or in the pores of the sorbent, thus creating a diffusive barrier to further adsorption of CO$_2$. In order to prevent this diffusive barrier of non-CO$_2$ gases from building up, two possible approaches have been suggested. One is to use a fan or blower to continually circulate fresh Martian atmosphere through the sorbent bed and then vented. As the incoming Martian atmosphere enters the sorption compressor, the residual permanent gases are displaced and the diffusive barrier is eliminated. An alternative method has been proposed that uses a small tank of high-pressure carbon dioxide to periodically send a purge of high-pressure gas through the sorption bed to flush out the permanent gases to the environment. This remains a potential challenge for sorption compressors.

(3) Efficient Daytime Heating—To minimize input power requirements, the sorbent bed must be thermally isolated from the radiator during the day. A very careful thermal design is likely to be required using a vacuum jacket and thermal isolators, and a thermal switch between the sorbent bed and the external radiator. A network of thermal conductors within the bed will assure reasonable uniformity of bed temperatures. This is a non-trivial challenge.

(4) Nighttime Cooling—The highly insulated adsorption compressor must cool down rapidly at night. A thermal switch allows good heat transfer from the network of thermal conductors within the bed to a radiator at night while minimizing heat loss from the sorbent bed during the day. As stated previously, this is a non-trivial requirement.

(5) Dust Filtration—Of primary concern is the avoidance of dust entering the sorption bed from the atmosphere. A full-scale plant will draw in a large amount of Martian atmosphere in 500–600 days, and this atmosphere will pass through valves, sensors, the sorbent bed and finally the reaction cell, all of which could possibly be degraded by dust impingement. Some approximate estimates of dust distributions are available but such distributions are bound to be highly variable with location, season, and dust storm activity. Passive dust filtration on Mars is made difficult by the low ambient pressures that requires a low pressure drop across the filter media.

(6) Pressure Drop Through Sorption Bed—To reduce the pressure drop through the sorption bed, one might arrange gas flow to be radial, not longitudinal. However, it might be simpler to assure that flow resistance through the bed is constant across its cross section with longitudinal flow. If the flow resistance is not uniform, attempts to flush out unadsorbed $Ar + N_2$ might not be effective.

JPL and Lockheed Martin Astronautics (LMA) developed a Mars Atmosphere Acquisition and Compression device (MAAC) (Rapp et al. 1997). JPL funded LMA to design, build and carry out tests on MAAC that could store at least 3 kg/day of CO_2.

An inner chamber contained a network of fins in which the Zeolite sorbent was placed, with a thermal transfer bar for transferring heat from the sorbent bed to an external radiator at night. An outer chamber was separated from the inner chamber by a vacuum jacket to minimize heat loss, and inlets were provided at either end. Flow was down the center tube and radially outward through the bed. A small fan was used to gently circulate gases to prevent build-up of non-adsorbed permanent gases such as Ar and N_2.

Initial tests were conducted on this unit with pure CO_2 and it worked very well. Three separate runs were made and the bed (20.5 kg of 13X Zeolite) stored 3.0, 3.3 and 3.4 kg of CO_2 in these runs, for storage mass percentages of 14.6, 16.0 and 16.6%.

When this system was tested using a Mars gas mixture, the performance suffered considerably. The total capacity of the sorbent bed dropped to about one third of its value in pure CO_2 and the rate of adsorption was inhibited. Clearly, the $Ar + N_2$ inhibited CO_2 adsorption. Apparently, the fan was not fully effective at flushing out unadsorbed $Ar + N_2$. This problem was not addressed because NASA funding dried up.

Another approach to a sorption compressor was developed originally by Brooks et al. (2005) utilizing microchannel structures to support thin layers of sorbent. The advantage of this approach is that the cycle time for heating and cooling is greatly reduced. However, the capacity of a single cell is low and this necessitates a complex array of many cells to produce a significant flow rate of compressed CO_2.

Merrell et al. (2007) claimed that a "microchannel CO$_2$ feed compressor, filter and sorption pump system, requiring 1080 g-CO$_2$/h at 100 kPa produced from a 600 Pa Martian atmosphere, are: 16.7 kg, 10.0 L, 1589 W cooling and heating and 61 W electrical power". Exactly how they arrived at these estimates was not clear.

Linne et al. (2013) reported further details on this approach.

None of these studies appear to have actually tested such a device within a chamber containing simulated Mars atmosphere during diurnal cycles. It is difficult at this stage to evaluate the merits of this approach.

1.3.1.2 Cryogenic Compressor

A pioneering method for compressing carbon dioxide in the Martian atmosphere was developed and demonstrated by Larry Clark at LMA and demonstrated at a breadboard level (Clark and Payne 2000; Clark 2003).

This process showed good potential for low-power high compression ratio system. The concept used a cryocooler to chill a surface below the freezing point of CO$_2$ and solid material collects on a metallic matrix. An insulated pressure vessel surrounds the cold surface and is valved off when the target volume of frozen CO$_2$ is collected. The frozen CO$_2$ is warmed up to ambient temperatures, producing a high-pressure CO$_2$ supply. During the freezing stage, a bypass flow is established with a low-power blower to purge non-condensable gases from the freeze chamber.

The energy required to freeze CO$_2$ includes cooling the gas to the freezing temperature and the heat of sublimation. Clark et al. (2001) performed the following estimate: Using an average incoming gas temperature of 210 K, cooling the gas to the freezing point requires 29 J/g and heat of sublimation is approximately 598 J/g for a total of 627 J/g. Cryocoolers can offer a Watt of cooling power for five Watts of electrical power, leading to an ideal power requirement of 0.871 W-h/g for acquiring solid CO$_2$. To produce say, $\sim$3 kg of CO$_2$ per day, requires 2.6 kW-h (about 400 W for 6 h). Vaporizing the CO$_2$ can be readily accomplished using the relative warmth of the Martian environment. Liquid forms and drips into the vessel as the CO$_2$ mass warms to the triple point, resulting in a saturated liquid at or near ambient Mars temperatures. This pure CO$_2$ supply can be maintained at these high delivery pressures by adding sufficient surface area on the vessel walls in the form of fins or support structure to supply enough heat during the outflow phase. These additional fins do not affect the acquisition process since the CO$_2$ collects on an internal heat exchanger that is separated from the vessel walls.

Since the proof-of-concept demonstration by Clark et al. (2001) later studies by Muscatello et al. (2012, 2014) evaluated several designs for the cryogenic accumulation chamber. Their best design was able to capture $\sim$94 g of CO$_2$ per hour from Mars simulant gas and had a capture efficiency of $\sim$64% at a flow rate of 84,000 cm^3/h.

In the cryogenic operational sequence, repeated cycles of CO$_2$ accumulation and release are carried out. In each cycle, some CO$_2$ is frozen out in the accumulation chamber as it is exposed to the Mars atmosphere, effectively acting as a cryopump. After closing the inlet to the atmosphere, the accumulated "dry ice" is later warmed up to regenerate pressurized gaseous CO_2. If configured in a tandem system, this gas can be fed directly into the SOXE while a second cryopump condenses the next batch of CO_2. In a single-pump system, heating of the condensed CO_2 compresses the gas to high pressure for storage, until adequate high pressure CO_2 is stored. At that time, the high pressure CO_2 is metered to the chemical reactor at an intermediate pressure (~ 1 bar). If storage occurs at pressures above the triple point of CO_2, ~ 5.1 bar, the system needs to accommodate the possibility that liquid CO_2 will be produced as an intermediate product.

During the accumulation phase, non-condensible gases (about 4.5%) are also ingested with the CO_2. These gases increase in concentration as CO_2 solidifies out, producing a diffusion barrier for new entering CO_2 to reach the cold finger. To avoid this, the cryogenic approach requires continuous pumping of atmosphere through the accumulation chamber to remove non-condensible gases. This adds yet another electro-mechanical device to the system.

The energy efficiency of the cryogenic approach is limited not only by the thermodynamics of compression and expansion, cooling and heating, and changes of phase, but also by the parasitic heat transfer from the accumulation chamber to the cryocooler. Heat gains by the accumulation chamber will reduce the effective cooling power of the cryocooler.

While Muscatello et al. (2012) did not explicitly report the parasitic heat load in their tests, it can be estimated from their data. Their cryocooler dissipated 350 W_e and provided "34 W_{th} lift at 150 K" for a typical $\sim 10{:}1$ ratio. It takes ~ 685 J to cool down and solidify 1 gm CO_2, corresponding to 5.25 g of solid CO_2 per W_{th}–h of heat removed by the cryocooler. Without parasitics, a 34 W_{th} lift at 150 K would thus produce 179 g/h of solid CO_2, compared to the observed best performance of 94 g/h, implying that parasitics negated about half the cooling power of the cooler. Their parasitic heat load must have been about 17 W_{th}. They also reported a 2/3 capture efficiency (fraction of entering CO_2 that was condensed).

To evaluate the feasibility of a cryogenic CAC for the MOXIE Project, JPL modeled two possible accumulation chambers as shown schematically in Fig. 1.5, in the context of the system flow diagram shown in Fig. 1.6. The minimal design used a copper mesh surrounded by an "air" space in a spherical enclosure. The high performance design utilized a copper mesh surrounded by a vacuum jacket, which was surrounded by insulation.

Using the minimal design, a spherical chamber was chosen with a storage capacity of 8.9 g of CO_2, to be captured in 7.5 cc of metal foam. A Ricor K543 cryocooler was assumed with nominal cooling power 2.8 W_{th} for lifting heat from 150 to 293 K. Considering only radiative and conductive parasitic heat gains, it was estimated that the net cooling rate (after subtracting parasitic heat gains) was 1.83 W_{th}. To prevent buildup of non-condensible gases Ar and N_2, which comprise about 4.5% of the Martian atmosphere, it is necessary to vent some of the cooling gas. This involved chilling the inert gases and releasing them to the atmosphere, thus reducing the net effective cooling power for producing solid CO_2 to about 1.67 W_{th}. With this cooling rate, the CO_2 accumulation rate

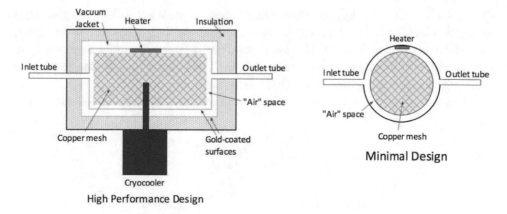

Fig. 1.5 Schematic cryogenic accumulation chamber

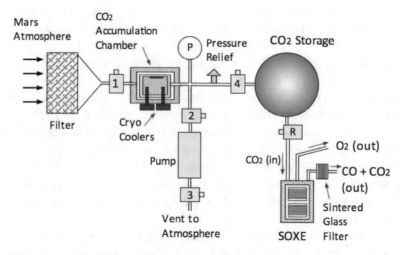

Fig. 1.6 MOXIE gas flow path with cryogenic CO$_2$ accumulation and compression

was estimated to be 8.9 g/h. Using the high performance design, the CO$_2$ accumulation rate was estimated to be 11.7 g/h.

For purposes of this analysis, it will be assumed that we need to store 55 g of pressurized CO$_2$ in order to produce 10 g/h of O$_2$ for 30 min. We roughly conjecture a series of 6 sequential acquisition cycles, with accumulation of roughly 8.9 g of CO$_2$ per cycle, thus storing about $(6 \times 8.9) \sim 55$ g of CO$_2$ in a storage tank.

The first issue is how many g/h of CO$_2$ can we store per accumulation cycle with a cryocooler? We assume use of a Ricor K543 with cooling power 2.8 W$_{th}$ to lift heat from 150 to 293 K ambient.

We begin by considering the simple CO$_2$ accumulator shown on the right side of Fig. 1.5. We choose a 7.5 cc spherical metal foam accumulator (85% porosity storing

CO_2 at 1.4 g/cc CO_2 density) and when fully saturated, it holds 8.9 g CO_2. The pressure is assumed to be 6 Torr. The emissivity of the metal foam and CO_2 is assumed to be 1.0 and the emissivity of the walls at 293 K is assumed to be 0.1. The radius of the foam is 1.2 cm and the radius of the containing sphere is 2.2 cm. This is a very simple design with the walls of the chamber exposed to a 293 K environment.

Radiative and conductive parasitic heat gain powers are calculated as follows:

$$P_{rad} = \frac{A_{cold}\sigma\left(T_h^4 - T_c^4\right)}{\frac{1}{\epsilon_{cold}} + \left(\frac{A_{cold}}{A_{hot}}\right)\left(\frac{1-\epsilon_{hot}}{\epsilon_{hot}}\right)} = 0.195\,\text{W}$$

$$P_{cond} = 4\pi\,k(T_h - T_c)\frac{R_1 R_2}{R_2 - R_1} = 0.774\,\text{W}$$

Thus the net cooling power at 150 K is reduced to 1.83 W_{th}.

Incoming CO_2 must be cooled from 293 to 150 K and frozen out at that temperature. Using the specific heat of CO_2 ($\sim$0.8 J/g-K) and the latent heat of deposition (571 J/g), the total heat that must be removed from CO_2 to deposit 1 g of solid CO_2 is 114 + 571 = 685 J/g. If there were no inert gases along with the CO_2, we could accumulate (1.83 J/s)/(685 J/g) = 0.00,267 g/s = 9.6 g/h of solid CO_2. However, gas is constantly being pumped through the accumulation chamber in order to prevent a buildup of inert gases that would act as a diffusion barrier preventing CO_2 from reaching the foam matrix. It is expected that only about 60% of the CO_2 entering the accumulation chamber would actually become solidified. The other 40% is vented along with the 4.5% of inert gases. Cooling this vented gas to 150 K requires about 0.45 $\times$ 114 = 51 J per g of CO_2 accumulated. Hence the actual energy requirement per g of CO_2 accumulated is 685 + 51 = 736 J, and only 8.9 g/h of CO_2 can be accumulated when cooling of vented gas is taken into account.

The more sophisticated accumulation chamber with a vacuum jacket and gold interior surfaces (left side of Fig. 1.5) would lead to lower parasitic heat gains. An estimate made by JPL is that the parasitic heat load with a vacuum jacket would be reduced from 1.0 W_{th} to roughly 0.4 W_{th}. Thus the net cooling power would be 2.4 W_{th}. In this case the estimated CO_2 accumulation rate would be 11.7 g/h.

According to JPL design principles, large energy margins for cryogenic systems should be applied. At the time of conceptual design, the recommended margin is 50% and at PDR it drops to 45%. However, margin should only be applied to calculated radiative and conductive loads. Sensible heat and latent heat are physical properties of the gas and do not need margins; their related heat loads are not influenced by accumulator design changes.

In the case of the simple accumulator design without a vacuum jacket, the parasitic load would increase from 0.97 W_{th} to 1.45 W_{th}, reducing the cooling power from 1.83 W_{th} to 1.35 W_{th}. The CO_2 accumulation rate would drop from 8.9 to 6.0 g/h.

In the case of the accumulator with a vacuum jacket, the parasitic load would increase from 0.4 W$_{th}$ to 0.6 W$_{th}$, reducing the cooling power from 2.4 W$_{th}$ to 2.2 W$_{th}$. The CO$_2$ accumulation rate would drop from 11.7 to 10.7 g/h.

The power requirement for the Ricor 543 cooler is nominally listed as 45 W$_e$. No data are available for the power requirement for the blower to circulate gas through the accumulation chamber. A wild guess is 30 W$_e$. In addition, another $\sim$20 W$_e$ would be required for electronics and control. Thus, a rough guess for the power requirement for the operation of accumulating solid CO$_2$ is 95 W$_e$.

Using the simple accumulator without a vacuum jacket and JPL margins, each accumulation cycle would have a duration of (8.9 g)/(6.0 g/h) $\sim$ 1.5 h. Six cycles would be needed to accumulate 55 g of CO$_2$. The energy requirement is

$$6 \times 1.5\,\text{h} \times 95\,\text{W}_e = 855\,\text{We-h}.$$

Using the vacuum jacket and JPL margins, each accumulation cycle would have a duration of (8.9 g)/(10.7 g/h) $\sim$ 0.83 h. Six cycles would require 5 h of operation. The energy requirement is 5 h $\times$ 95 W$_e$ = 475 W$_e$-h.

The energy required to warm the solid CO$_2$ to pressurized gaseous CO$_2$ is not included here since the electric power required to vaporize and warm 55 g of CO$_2$ is only 685 $\times$ 55 W$_e$-s = 10 W$_e$-h.

1.3.1.3 Mechanical Compressor Approach

The potential benefits of a mechanical compressor for our application include much lower total energy consumption per oxygen production cycle, and a significant reduction in system complexity (including elimination of valves, and direct straight-through operation without batch cycling). This system is shown schematically in Fig. 1.7.

Of the available mechanical compression technologies, the scroll pump (or equivalently, compressor) was chosen by the MOXIE Project[4] for this application. This type of pump takes in a fixed volume of gas with each rotation and compresses it to a substantially smaller fixed volume. It does this by rotating a movable scroll within a fixed second scroll. Thus, for a particular rotation rate, and assuming a constant volumetric efficiency, the output mass flow rate will theoretically scale linearly with the input gas density. The scrolls are tipped with a Teflon-like material that produces a tip seal. The power requirement is due to the thermodynamics of physical compression, the motor resistance, and the friction in the tip seals and the pump bearings. The power required for the motor and pump bearings seems to be fairly well fixed. However, the power dissipated by tip seal friction can vary, depending on how the seals are adjusted. Shims are used to adjust the tip seal gap to minimize friction while maintaining adequate seals. If the tip seals are too loose, the power is reduced but the flow is non-linear. If the tip seals are too tight, the

---

[4]https://mars.nasa.gov/mars2020/mission/instruments/moxie/; https://ssed.gsfc.nasa.gov/IPM/PDF/1134.pdf.

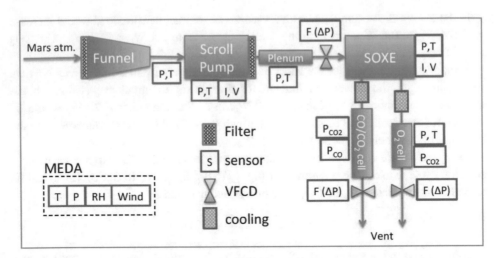

Fig. 1.7 In a completely valve-free design, the MOXIE scroll compressor pulls in filtered Martian air and exhausts the compressed air through a second filter into a plenum, which provides a low pass filter against pressure variations. This gas is fed into the SOXE, which produces pure O_2 and exhausts the CO product combined with unreacted CO_2 and any trace atmospheric components. Viscous Flow Control Devices (VFCD) allow the pressure to be regulated by adjusting the pump motor speed

power requirement can increase markedly. When they are optimally adjusted, the power requirement is minimized and the dependence of mass throughput versus pump speed is expected to be fairly linear over a wide range of pump speeds.

The MOXIE flight experiment adopted a scroll pump as its compressor for the Mars atmosphere.

The scroll compressor for MOXIE is under development by Air Squared, Inc., of Denver, CO. The scroll pump is illustrated in Fig. 1.8. The power consumed by the pump rises linearly with pump speed. So does the volumetric flow rate. Since development of the pump is still a work in progress, it is difficult at this time to pin down the power requirement for any given flow rate. Preliminary modeling suggested that power requirements break down roughly as shown below. However recent testing in late 2017 suggests these estimates are too low.

$$\text{Motor} = 18\,\text{W}$$
$$\text{Friction in the tip seals} = 7\,\text{W}$$
$$\text{Pump bearings} = 11\,\text{W}$$
$$\text{Thermodynamic compression} = X(\text{W})$$
$$\text{Total power} = 36\,\text{W} + X$$

I have denoted the power used for thermodynamic compression as "X" because it needs further discussion.

Fig. 1.8 A scroll pump
compresses gas by means of an
orbital motion of one set of
involutes against a fixed set.
A sliding PTFE-based "tip
seal" prevents gas from leaking
under the ends of the scrolls

The scroll pump takes in a suction volume of V_1 = 30.1 cc. The scroll pump compresses each suction volume by a factor of 5.8:1, so the delivered volume to the downstream plenum is V_2 = 5.19 cc per suction volume. Assuming that the Mars pressure is 5.25 Torr (average for Jezero Crater) the pressure in each 5.19 cc volume can be calculated for adiabatic compression:

$$P_1 V_1{}^g = P_2 V_2{}^g$$

where $g = c_p/c_v \sim 1.28$

Thus we calculate $P_2 = 5.25 \times (5.8)^{1.28} = 49.8$ Torr.

Let the pressure in the downstream plenum be denoted P_3 (typically several hundred Torr). As the scroll pump rotates, the gas in the 5.19 cc volume at 49.8 Torr is exposed to the downstream plenum at P_3 and gas "backwashes" into each 5.19 cc volume bringing the pressure up to slightly less than P_3 (the plenum volume is large compared to V_2). As the scroll pump closes its connection to the downstream plenum, the volume V_2 at P_3 must be pushed into the plenum (V_2 disappears). This requires additional work.

Thus we can model the thermodynamic power needed to operate the scroll pump as a series of three steps:

(1) Adiabatic compression of suction volume (30.1 cc) from 5.25 Torr to 49.8 Torr.
(2) Backfill of gas at P_3 into two 5.19 cc volumes raising the pressure to almost P_3 at constant volume—no work done.
(3) Pushing the gas at P_3 in two 5.19 cc volumes into the plenum. Work = $P_3 \Delta V$.

The work required for the first step (adiabatic expansion to (1/5.8) times the suction volume) is:

$$W_1 = [1/(g-1)]P_1V_1\left\{[V_2/V_1]^{(1-g)} - 1\right\}$$
$$W_1 = [1/0.28](5.25)(30.1)\left\{(0.1724)^{(-0.28)} - 1\right\}$$
$$W_1 = 360(cc - Torr)$$
$$\text{with } P_1 = 5.25 \, Torr.$$

The second step merely involves "pushing" a volume $V_2 = 5.19$ cc into a plenum at pressure P_3. This work is simply (5.19) (P_3)

If P_3 is 750 Torr, the second step requires $W_2 = 3890$ (cc-Torr).

One can see that the vast majority of work done in compression is not the first step, but rather the second step. The overall work required per cycle is $360 + 3890 = 4250$ cc-Torr.

If P_3 were reduced to say, 150 Torr, the work involved in the second step would be reduced to (5.19) $(150) = 780$ cc-Torr, and the total work would be reduced to 1140 cc-Torr.

At 150 Torr, total work for compression is reduced to about 27% of that at 750 Torr. We can convert these work figures to Joules using:

$$1 \, cc\text{-}Torr = 1/750 \, cc\text{-}bar = 0.1/750 \, J = 0.000133 \, J$$

Next we convert these work figures to power by noting that the duration of a pump rotation at a rotational speed of 3500 RPM, is

$$T = 60/3500 = 0.0171 \, s$$

Thus to convert work in arbitrary units to J/s (Watts) we multiply cc-Torr by:

$$0.000133/0.0171 = 0.0078$$

We may conclude that the power for the first stage of compression is

$$360 \times 0.0078 = 2.8 \, W$$

The power for the second stage is

$$4250 \times 0.0078 = 33 \, W \, @ \, 750 \, Torr$$
$$780 \times 0.0078 = 6 \, W \, @ \, 150 \, Torr$$

We find the power levels to be

$$X_1 = 2.8\,\text{W}$$
$$X_2 = 33\,\text{W} @ 750\,\text{Torr}$$
$$X_2 = 13\,\text{W} @ 300\,\text{Torr}$$
$$X_2 = 6\,\text{W} @ 150\,\text{Torr}$$

The total power to operate the pump at 3500 RPM and 5.25 Torr inlet pressure is estimated to be ~ 110 W at 750 Torr, 76 W at 300 Torr, and ~ 63 W at 150 Torr. However, recent test data suggest that actual power requirements might be somewhat higher.

There seems to be little doubt that going beyond MOXIE to scaled-up follow-on demonstrations, operating at a pressure in the range 100 Torr to 300 Torr would pay significant dividends in lower power consumption and possibly more uniform pressure output than operating at higher pressures such as 750 Torr. This is also true for MOXIE.

These pumps can be made in single-stage and two-stage models. The two-stage models have significant advantages but have yet to be adequately tested.

Since the scroll pump is a volumetric multiplying device, we can assume that the throughput of the pump into a 1 bar reservoir is proportional to the inlet density. We can estimate the variability of inlet conditions for various seasons and times of day. For purposes of illustration we choose a nominal 5.25 Torr yearly average pressure corresponding to a relatively high-pressure region on Mars. In this case the inlet gas density varies from

Lowest density: $11.5\,\text{g/m}^3$ at 280 K and 4.5 Torr

Highest density: $24.5\,\text{g/m}^3$ at 170 K and 6.0 Torr.

This represents a dynamic range of 24.5/11.5; or more than a factor of two in density. A very preliminary specification for the MOXIE scroll pump was that at 3500 RPM, it would pump about 83 g/h to a 1 bar reservoir with inlet conditions at 7.0 Torr, 298 K, where the density is $16.6\,\text{g/m}^3$.

Therefore, the estimated capabilities for pumping speed of the scroll pump at the highest and lowest Mars inlet densities are:

Lowest density: $11.5/16.6 \times 83 = 58\,\text{g/h}$

Highest density: $24.5/16.6 \times 83 = 122\,\text{g/h}$.

The above estimates do not take into account the pressure drop in the HEPA filter interposed between the atmosphere and the scroll pump. Furthermore, the density of the Mars gas might decrease in the inlet tube if the Mars gas is warmed by heat conduction through the tube walls.

The MOXIE Project utilizes a single-stage scroll pump developed from a COTS commercial model pump for expediency. An optimal system might utilize more than one two-stage pumps in series. Alternatively, other types of mechanical pumps might offer benefits. This has yet to be investigated.

1.3.2 Dust Rejection

1.3.2.1 Introduction

The most obvious and simplest approach to reject dust from entering the reactor system is to use a HEPA filter. The clean filter will introduce a small pressure drop, thus reducing the effective inlet pressure to the scroll pump.

HEPA filters utilize folded filter material with typically about twenty times the filter area compared to the face area. As dust accumulates on the filter through use, the dust will fill the pores of the filter material, introducing additional pressure drop. The relationship between pressure drop and dust loading under Martian conditions remains uncertain.

Most research on HEPA filters was done at ~ 1 bar, and applicability to Martian conditions is uncertain. Most experimental correlations were given in terms of dust loads (g/m^2). However, it is not the mass of the dust that is most important, but rather, the blocking area of the dust particles. The blocking area of 1 g/m^2 of dust particles is ten times greater for particles 1/10 the radius. Thus we see that the particle size distribution is of great importance, as well as the dust mass loading.

1.3.2.2 Physical Properties of Mars Dust

A number of published articles utilized optical measurements on Mars to infer properties of dust suspended in the "air". A widely accepted result is that of Tomasco et al. (1999) who estimated the area-weighted average radius of a Mars dust particle to be 1.6 μm. The distribution of particle sizes found by Tomasco et al. was a gamma function that is unbounded at small radii. This leads to problems in using the function to estimate averages. Hecht et al. (2017) showed that the cumulative volume function for Mars regolith at the Phoenix site dropped sharply with decreasing particle size, indicating that the number of particles at smaller sizes does not rise without limit, but rather, flattens out.

Deirmendjian (1964) had suggested a modified function that neglects very small particles. Using the modified gamma function for the distribution of particle sizes proposed by Deirmendjian (1964) we arrive at the distributions shown in Fig. 1.9. The curves are drawn for a most probable radius $r_m \sim 0.25$ μm. The average radius is about 0.66 μm, the average radius of an area-weighted distribution is 1.6 μm, and the average radius of a mass-weighted distribution is 2.5 μm. These measurements pertain to the full path length through the atmosphere and might not be representative of dust suspended a meter above the surface, which could vary with season, location and weather. We will use these estimates here, knowing that they are highly approximate.

To estimate the blocking area of a collection of dust particles, use the area-weighted average radius. To estimate the mass of an average dust particle, use the mass-weighted average radius.

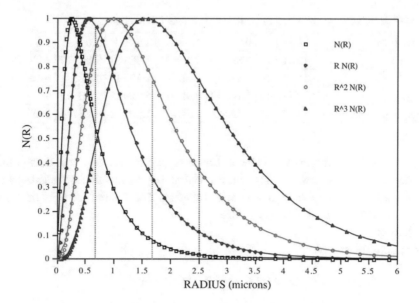

Fig. 1.9 Assumed distribution of particle sizes on Mars

The density of Mars dust particles depends on the porosity. Allen et al. (1998) estimated it to be roughly 1.5 g/cm^3 based on PF data. Based on these approximations, we calculate:

The area-weighted average area of a Mars dust particle is $A_d = \pi (1.6)^2$ μm$^2 = 8.0 \times 10^{-8}$ cm^2.

The volume-weighted average volume of a Mars dust particle is $V_d = 4/3 \ \pi (2.5)^3$ μm$^3 = 6.6 \times 10^{-11}$ cm^3.

The mass-weighted average mass of a Mars dust particle is roughly estimated to be: $M_d = 1.5 \times 6.6 \times 10^{-11} = 1.0 \times 10^{-10}$ g.

The area, volume and mass of particles of different diameter are given in Table 1.9.

1.3.2.3   Optical Properties of Mars Dust

Let A_s be the effective cross sectional area of a particle for light scattering and absorption. A_s is related to the geometrical area of a particle, A_d, by the relation:

$$A_s = A_d Q_{ext}$$

where Q_{ext} is the so-called extinction efficiency or scattering efficiency. It turns out that because of diffraction effects, Q_{ext} is >1. Pollack's 1990 model had $Q_{ext} = 2.74$, but a more recent model by Tomasko (1999) based on PF data set $Q_{ext} = 2.6$.

In addition, most of the scattering is in the forward direction. We must distinguish between optical properties for imaging and optical properties of dust laying on a solar

Table 1.9 Area, volume and mass of particles of different diameter

Diameter (μm)	0.1	0.3	1.0	2.0	3.0	5.0
Area (cm^2)	3.1E−10	2.8E−09	3.1E−08	1.3E−07	2.8E−07	7.9E−07
Volume (cm^3)	5.2E−16	1.4E−14	5.2E−13	4.2E−12	1.4E−11	6.5E−11
Mass (g)	7.9E−16	2.1E−14	7.9E−13	6.3E−12	2.1E−11	9.8E−11
Number of particles for 1 g	1.3E+15	4.7E+13	1.3E+12	1.6E+11	4.7E+10	1.0E+10

array. For imaging, what matters is the effective scattering area A_s, even though most of the scattering is in the forward direction. For estimating degradation of solar cells due to dust, forward scattering acts as if it is non-scattering. The fraction of light that is either absorbed or scattered backwards is ~ 0.23.

Therefore:

For optical transmission,

$$A_s = 2.6\,A_d = 2.1 \times 10^{-7}\,cm^2.$$

For obfuscation of solar cells:

$$A_s = 0.23 \times 2.6\ A_d = 0.60\,A_d = 5.0 \times 10^{-8} cm^2.$$

1.3.2.4 Optical Depth on Mars

The definition of optical depth is

D = (N = number of particles in vertical column) x (As = average area for scattering and absorption per particle)/(cross sectional area of column)

For a column of cross sectional area 1 cm^2:

$$N = (D)/(A_s)$$

The optical depth represents the fractional blocking area when "seeing" through a vertical column of Mars atmosphere. See Fig. 1.10.

There are now two questions.

The first question is which value of A_s should be used (imaging or solar cell obfuscation)? Measurement of optical depth is described by Petrova et al. (2011). It appears that the best value of A_s is that corresponding to total brightness; hence we use the value for obfuscation of solar cells:

$$A_s = 0.60\,A_p$$

Fig. 1.10 Projection of view through a tube showing dust particles

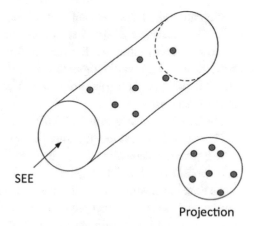

SEE

Projection

As Petrova et al. (2011) pointed out, "Mars commonly has an optical depth of 0.3–0.7". Rover measurements typically find D ∼ 0.5 in "clear" weather. HiRise measurements indicated a range of 0.43–0.53.

For our purposes, we use D ∼ 0.5 for "clear" weather (D can reach values as high as 4 in global dust storms).

1.3.2.5   Dust Particles Per Unit Volume in Mars Atmosphere

The second question is: What is the mixing height for dust in the Mars atmosphere? Heavens et al. (2011) found this to be highly variable in the range 15–25 km. Landis (1996) used 20 km. Here, we use 20 km.

The number of particles in a column 20 km high by 1 cm^2 cross sectional area is estimated to be:

$$N = (D)/(A_s) = 0.6/8.0 \times 10^{-8} = 7.5 \times 10^6 \text{ particles}$$

The volume of the column is $1 \times 2 \times 10^6 \text{ cm}^3 = 2 \times 10^6 \text{ cm}^3$

The number of particles per cm^3 is then

N ≈ 4 dust particles per cm^3

As a "sanity check" on the conclusion there are 4 dust particles per cm^3, suppose we are an observer on Mars and we look across a length L of Mars atmosphere at surroundings.

If we look across the Mars atmosphere through a tube of cross sectional area 1 cm^2 and length L, we encounter an imaging blockage of 4 dust particles per cm^3 × L cm^3 $2.1 \times 10^{-7} \text{ cm}^2/1 \text{ cm}^2$.

For L = 100 m = 10^4 cm, the imaging blockage is 0.008.

For L = 1 km, the imaging blockage is 0.08 = 8%.

Pictures of Mars probably fit this reasonably well. Figure 1.11 suggests such levels of obfuscation.

We can make a second sanity check on the conclusion there are 4 dust particles per cm^3, based on degradation of solar cells on Mars.

Landis (1996) estimated the settling time for a 1 cm^2 column of dust on Mars to be about 80 sols. That would imply that 6×10^6 particles would settle in 80 sols, or the daily rate of deposition on a solar array would be $6 \times 10^6/80 = 75,000$ dust particles/sol-cm^2 settle on an array.

The blocking area would be $75,000 \times 8.0 \times 10^{-8}$ $cm^2 = 0.006$ cm^2 per sol.

That would suggest that the initial degradation of the solar cell would be 0.6% per sol.

The mass deposition of dust is estimated to be $75,000 \times 1.0 \times 10^{-10}/$ $(3600 \times 24) = 8.7 \times 10^{-11}$ g/cm^2-s $= 8.7 \times 10^{-7}$ g/m^2-s

As time progresses, dust will settle on a solar array, but as dust builds up, some will be removed. If we assume that there is a steady addition of 75,000 particles per sol, but the rate of removal is proportional to the fractional dust area on the cells (F), we obtain the rate of buildup of obscuration is:

$$dF/dt = 0.006 - K\,F$$

where S = sols after exposure of array, F = fractional obscuration, K = unknown constant. Integrating from t = 0 to t = S:

$$F = (0.006/K)\,[1 - \exp(-K\,S)]$$

where F approaches 0.006/K for large S. Based on the observation that the array ultimately reaches $\sim 30\%$ coverage at long times, K would be roughly 0.02. The resultant curve of F versus S is roughly in agreement with data from Spirit and Opportunity (see Appendix B of Rapp 2015).

Fig. 1.11 View from Mars rover showing optical depth on a clear day

1.3.2.6   Potential Dust Intake

Dust intake is expected to derive from three main processes:

(1) During the landing process, retro rockets pointing downward stirring up dust and surface material.
(2) The intake sucks in Martian "air", and the "air" contains suspended dust (roughly estimated to be about 4 particles per cm^3). This occurs 24/7 for 14 months of operation during the lifetime of the MOXIE mission ($\sim 10,000$ h).
(3) Dust intake independent of pumping. Winds blow on Mars much of the time at up to 10 m/s. Winds will blow Martian "air" against the filter, possibly depositing some of the suspended particles. The suspended particles include suspended dust, saltated particles, particles stirred up during the landing process, and particles suspended by dust devils or dust storms. This also occurs 24/7 for 14 months of operation during the lifetime of the MOXIE mission ($\sim 10,000$ h).

As we showed in Sect. 1.1.5, the required intake of CO_2 for a crew of six for a process that utilizes $\sim 50\%$ of the CO_2 is about 19 kg/h. If some of the pressurized CO_2 in the electrolysis tail gas can be recovered and recirculated, this would be reduced. For a process that uses 95% of the incoming CO_2, the requirement is for 10 kg/h. As a compromise, it will be assumed here for purposes of illustration, that the required flow of CO_2 is 14 kg/h. This corresponds to a pumped volumetric flow rate of roughly:

$$V = (14,000)/44)(82.01)(240)/(6/750) = 8 \times 10^8 cm^3/h$$

With 4 particles/cm^3, the dust loading due to pumping is estimated to be 3×10^9 dust particles/h. Over 10,000 h, total dust ingestion is 3×10^{13} particles. In Sect. 1.3.2.1 we estimated the mass-weighted average mass of a dust particle to be about 1×10^{-10} g, so the total dust loading due to pumping is estimated to be about 3 kg.

The volume of ingested dust is roughly estimated to be $6.6 \times 10^{-11} \times 3 \times 10^{13} \sim 2000$ cm^3 = 2 L.

The total blocking area of the dust is roughly estimated to be 3×10^{13} particles $\times 8 \times 10^{-8}$ cm^2 = 3×10^5 cm^2 = 30 m^2.

All of the above deals only with Mars "air" sucked in through the filter by the pump. Additional deposition due to wind impact, dust devils, dust storms, etc. above and beyond that due to pumping action, must be added to this estimate. However, baffling can be used to reduce the intake from these sources of dust.

1.3.2.7   Dust Rejection Systems

Dust rejection systems will undoubtedly use some combination of electrostatics and physical filters. Use of a filter inevitably leads to a pressure drop across the filter. Since a typical atmospheric pressure on Mars is about 5 Torr, and the throughput of a pump is typically proportional to the pressure, any significant pressure drop across the filter will reduce the throughput. HEPA filters are likely choices for filters because they intercept dust particles in the size range encountered on Mars. However, even a clean HEPA filter has a small pressure drop. Almost all applications of HEPA filters have been at ~ 750 Torr pressure where a pressure drop of a fraction of a Torr or even a few Torr would be inconsequential. There is only sparse data on pressure drop across HEPA filters, and almost no data at Mars pressures. Recent experiments conducted by McLean et al. (2017) suggest that the pressure drop through a clean HEPA filter at Mars conditions at velocity of a few cm/s might be about 0.2 Torr.

A challenge for a filter is to provide enough surface area that the filter does not get clogged, and thereby increase the pressure drop above and beyond that of a clean filter. As in the case of clean filters, data for the effects of dust accumulation on pressure drop are sparse and difficult to interpret for our purposes.

As far as we can tell from a sparse literature on pressure drop versus dust loading, it seems likely that the pressure drop through a clean filter will probably be a few tenths of a Torr. It is difficult to predict what dust loadings would double the ΔP relative to the ΔP for a clean filter. A wild guess is that a dust loading of 1 g/m^2 of filter area might double the ΔP relative to the ΔP for a clean filter. Typical HEPA filters are folded so the actual area of filter is about 20 times the open face area. Hence, we guess that the dust loading of 20 g/m^2 of face area might double the ΔP relative to the ΔP for a clean filter. To spread 3 kg of dust over the face area of HEPA filters would require about 3000/20 = 150 m^2 of face area. However it must be emphasized that this estimate is very fragile and is based on flimsy evidence.

Since Mars dust particles are likely to be electrostatically charged, it will be necessary to validate that electrostatic self-cleaning processes work for both neutral and charged particles.

Calle et al. (2011) described an electrostatic precipitator for Mars dust. As they put it:

With an atmospheric pressure between 7 and 10 mbars, the Martian atmosphere severely limits the potentials that can be applied to the electrodes of an electrostatic precipitator. An electrostatic precipitator collects dust particles that have been electrostatically charged by driving these charged dust particles onto one of the electrodes by means of an applied electric field. At 7–10 mbars, Townsend breakdown occurs at relatively low potentials.

The breakdown potential in Townsend breakdown in a uniform electric field depends on the product of the gap length d and the gas number density. This relationship is known as Paschen's law.

Calle et al. (2011) performed experiments with a premixed gas to emulate the Martian atmosphere at pressures around 9 Torr (roughly double that expected for Mars). For d $\sim$ 5 mm, the breakdown voltage was about 725 V.

An electrostatic precipitator consists of two electrodes set at an electrostatic potential difference that can drive charged particles to one of the electrodes for collection. There are two general types of electrostatic precipitators: Single-stage precipitators, where particle charging and particle collection take place in a single stage; and two-stage precipitators, with a pre-charging stage and a collecting stage. In both precipitator types, dust particles are charged using corona generation around the high voltage discharge electrode, which ionizes gas molecules. These ions are accelerated by the electric field in the region between the electrodes, but numerous collisions with gas molecules results in a constant drift velocity characterized. The ions transfer charge to dust particles encountered in their path as they drift to the grounded collecting electrode. The ions form a space charge that modifies the applied field between the electrodes.

M. K. Mazumder has carried out some very impressive research and development of electrostatic systems for dust rejection. However, almost all of his work addressed protecting surfaces such as solar arrays, mirrors and radiator surfaces. Some of this will be important for ISRU systems, but does not seem to be immediately applicable to dust intake for pumping systems. Nevertheless, one can conceptually imagine that his systems might be adaptable to controlling dust intake into atmosphere acquisition ducts.

1.4   Processes Utilizing Mainly CO$_2$ from the Atmosphere

1.4.1   The Reverse Water-Gas Shift Reaction

The water-gas shift reaction is widely used by industry to convert relatively useless CO + H$_2$O into much more useful hydrogen. However, if reaction conditions are adjusted to reverse the reaction so it has the form:

$$CO_2 + H_2 \Rightarrow CO + H_2O \quad \text{(catalyst required)}$$

then CO$_2$ can be converted to water. If that water is electrolyzed:

$$2H_2O + \text{electricity} \Rightarrow 2H_2 + O_2$$

the net effect of the two reactions is conversion of CO$_2$ to O$_2$. Ideally, all the hydrogen used in the first reaction is regenerated in the electrolysis reaction, so no net hydrogen is required. Actually, some hydrogen will probably be lost if the first reaction does not go to completion, although use of a hydrogen recovery membrane can minimize this loss. The above two reactions in concert represent what is referred to as the "reverse water-gas shift" (RWGS) approach to ISRU. The RWGS approach was introduced and championed by Robert Zubrin of Pioneer Astronautics (Zubrin et al. 1997, 1998).

Note that the reagents for the RWGS reaction are the same as the reagents for the Sabatier/Electrolysis (S/E) reaction (see Sect. 1.5). The main difference (aside from use of a different catalyst) is that the S/E process has a favorable equilibrium at lower temperatures (200–300 °C) while the RWGS has a more favorable equilibrium at much higher temperatures (>600 °C). If one considers the combined equilibria where catalysts are present that allow both reactions to take place, the S/E process will be dominant below about 400 °C, and the reaction products will be mainly CH$_4$ + 2H$_2$O. At temperatures above about 650 °C, methane production falls off to nil and the RWGS products (CO + H$_2$O) are dominant. Between about 400 and 650 °C, a transition zone exists, where both reactions take place. In this zone, CO production rapidly rises as the temperature increases from 400 to 650 °C while methane production falls sharply over this temperature range. However, no matter how high the temperature is raised, roughly half of the $CO_2 + H_2$ remains unreacted in the high temperature regime.

Figure 1.12 shows reagent and product flow rates (assuming equilibrium is attained) for various chemical species when 44 mg/s of carbon dioxide and 2 mg/s of hydrogen (1 mmol of each) are introduced into a reactor at any temperature (note that the carbon dioxide flow rate is divided by 2 before plotting to reduce the height of the vertical scale) and both the RWGS and S/E reactions are catalyzed to equilibrium in the reactor. It should be noted that this is the proper stoichiometric ratio for the RWGS reaction, but represents an excess of CO_2 of a factor of 4 for the S/E process. With this excess of CO_2, essentially all of the hydrogen will be reacted in those regimes where the S/E process takes place with high yield.

At lower temperatures (200–300 °C), equilibrium would dictate that almost all the hydrogen is used up to produce methane and water, and the excess carbon dioxide is depleted by an equivalent amount. This is the S/E region. Almost no CO is formed. By contrast, at high temperatures (>650 °C) CO and H_2O are the principal products, and very little methane is formed, but roughly half of the initial carbon dioxide and hydrogen remains unreacted in the product stream. This is the RWGS region.

Zubrin et al. (1997; 1998) used copper on alumina catalysts with which only the RWGS reaction takes place so that the S/E reaction does not occur in the reactor. Then assuming equilibrium for the RWGS reaction, and no S/E reaction occurs, Zubrin pointed out that one could operate the RWGS at lower temperatures. For example, at 400 °C the equilibrium constant for the RWGS reaction is 0.1. In a simple reactor at 400 °C, the pressures satisfy the equilibrium relationship:

$$K = \{p(H_2O)\,p(CO)\}/\{p(CO_2)\,p(H_2)\} = 0.1$$

Fig. 1.12 Product flow rates (assuming equilibrium is attained) for various chemical species when 44 mg/s of carbon dioxide and 2 mg/s of hydrogen are introduced into a reactor at any temperature. Note that the actual carbon dioxide flow rate is divided by 2 before plotting in order to reduce the range of the vertical scale

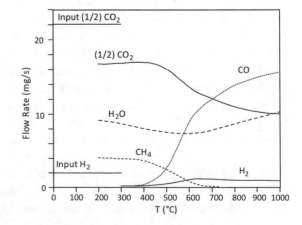

For a stoichiometric mixture, if we put $p(CO_2) = p(H_2) = x$ and $p(H_2O) = p(CO) = 1 -x$, at a total pressure of 2 bar, partial pressures are determined by

$$x2/(1 - x)\,2 = 0.1.$$

We find:

$$p(CO_2) = p(H_2) = 0.76 \, bar$$
$$p(H_2O) = p(CO) = 0.24 \, bar$$

The conversion of $\{CO_2 + H_2\}$ to $\{H_2O + CO\}$ at equilibrium is only 24%. For every mole of water produced, roughly 3 mol of hydrogen remains unreacted. If the spent gases are vented, and the water is condensed out, ~ 3 mol of hydrogen will be vented in producing a mole of water. This mole of water, in turn, will produce 0.5 mol of oxygen by electrolysis, if there are no losses. Thus it would takes 6 mol of hydrogen to produce one mole of oxygen by such a RWGS process at 400 °C. This is considerably more than the 2 mol of hydrogen needed to produce 1 mol of oxygen by the S/E process.

Zubrin et al. (1997) suggested that there are a number of ways to drive the equilibrium to the right:

(a) Overload the reactor with CO_2 to force almost complete consumption of H_2, and then separate excess CO_2 and recycle the excess CO_2 in the exhaust stream back into the reactor.
(b) Overload the reactor with excess H_2 to force almost complete consumption of CO_2, and then separate excess H_2 and recycle the excess H_2 in the exhaust stream back into the reactor.
(c) Remove water vapor from the reactor using a condenser or a desiccant, driving the equilibrium to the right.
(d) Combine (a) and (c).
(e) Combine (b) and (c).

The approach for separation and recycling suggested by Zubrin et al. (1997) {(a) or (b) above} involves use of a special membrane that transmits only the desired component, followed by a compressor to compress the component up to reactor entry pressures. This adds considerable complexity. There are also concerns that membranes may not be effective or may be poisoned by other constituents of the reactor efflux. (Conventional Nafion membrane systems are poisoned by CO). However, Zubrin et al. (1997) used such a membrane in their work.

Approach (c) above implies that in order to drive the RWGS reaction to the right (CO + H_2O), water produced by the RWGS reaction must be condensed out downstream of the reactor and the gases recirculated with continuous admixing of a smaller flow of feed gases. This is illustrated schematically in Fig. 1.13.

An appropriate molar flow rate (A) of feed gases continuously enters a RWGS reactor such that an appropriate amount of product water is continuously produced. In addition, a much higher flow rate (B ≫ A) of recirculated gases is also introduced to the reactor in parallel. Equivalent molar rates (A) of spent gas and water are continuously removed from the condenser. Since water produced by the reaction is continuously swept out of the reactor by the relatively high flow rate and condensed in the condenser, the water concentration in the reactor in the steady state will be greatly reduced compared to a simple reactor. In fact, as the ratio B/A becomes large, the water pressure tends to approach the vapor pressure of water at the condenser temperature, and most of the gas in the reactor is CO. For example, with a high rate of recirculation and continuous condensation of water, one might hope that the pressure of water in the reactor might be reduced to something like the vapor pressure of water at 10 °C, which is ~0.01 bar. In this case, the equilibrium relation (in bars) becomes:

$$K = \{(0.01)p(CO)\}/\{p(CO_2)p(H_2)\} = 0.1$$

If a stoichiometric mixture is used, we set $p(CO_2) = p(H_2) = "x"$ and we find that

$$p(CO) = 10\,p(CO_2)p(H_2) = 10\,x^2$$

At any total reactor pressure P_t (bars), we can write

$$p(CO) \sim (P_t - 2x)$$
$$p(CO2) = p(H_2) = x$$

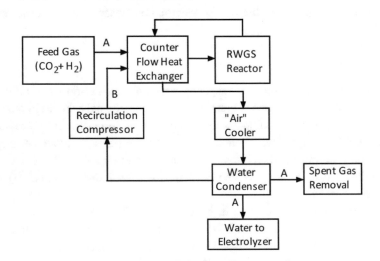

Fig. 1.13 RWGS system with recirculation

so that

$$(P_t - 2x) = 10\,x^2$$

and if $P_t = 2$ bar, we find

$$p(CO_2) = p(H_2) = 0.36\,bar$$
$$p(CO) = 1.28\,bar$$

The conversion of [CO$_2$ + H$_2$] to [H$_2$O + CO] is now increased to 1.28/1.64 = 78%. With a 78% conversion efficiency, this implies that for every mole of water produced, 0.22 mol of hydrogen will be vented. As before, one mole of water produces 0.5 mol oxygen by electrolysis, so the hydrogen required to produce 1 mol of oxygen is reduced from ~6 mol without recirculation to 1/0.78 = 1.3 mol with recirculation. This is considerably less than the two moles of hydrogen required to produce a mole of oxygen by the S/E process.

Note that if P_t is increased to say, 10 bar, x ~ 0.9 and (P$_t$ − 2x) ~ 8.2 so the conversion to CO is increased to about 90%. With a 90% conversion efficiency, this implies that for every mole of water produced, 0.11 mol of hydrogen would be vented if there were no hydrogen recovery system. As before, one mole of water produces 0.5 mol oxygen by electrolysis, so the hydrogen required to produce 1 mol of oxygen is reduced to 0.22 mol with recirculation at 10 bars even without hydrogen recovery. This is roughly 1/10 the hydrogen required by the S/E process.

The price paid for the recirculation process includes (i) the need for a recirculating compressor that must overcome the pressure drop of the reactor and the heat exchanger, and (ii) the heat loss in the heat exchanger due to the inevitable ΔT that prevents full recuperation of heat from the product stream to the inlet stream.

In summary, Zubrin and co-workers have proposed three approaches for increasing the hydrogen conversion efficiency of a RWGS reactor:

(i)   Water condensation to minimize water vapor pressure and recirculation of CO + CO$_2$.
(ii) Use of excess hydrogen (off-stoichiometric mixtures) to force the reaction to the right, with membrane recovery of unreacted hydrogen.
(iii) Increasing the reactor pressure.

In general, the conversion ratio can be calculated as follows. Let the total pressure be Pt and let the H$_2$/CO$_2$ molar feed ratio to the reactor be denoted as R. Let us also assume that the effective water vapor pressure in the reactor is maintained at ~0.01 bar by the condenser and recirculation. Then, the appropriate equilibrium relationship at 400 °C is determined as follows:

$$p(CO) \sim P_o\{x/(1+R)\}$$
$$p(CO_2) = P_o[(1-x)/(1+R)]$$
$$p(H_2) = P_o[R/(1+R)] - P_o[x/(1+R)]$$
$$p(H_2O) = 0.01 \text{ bar}$$
$$P_t = P_o - P_o x/(1+R)$$

where P_o is the total pressure that would prevail if no reaction took place. Thus:

$$x^2 - x\{(0.1/P_o + 1)(1+R)\} + R = 0$$

The results of solving this equation are given in Fig. 1.14. Several results are clear. Increasing the pressure improves the $CO_2 \Rightarrow CO$ conversion efficiency but the rate of improvement diminishes when the pressure exceeds about 3 bars. Increasing the H_2/CO_2 mixture ratio improves the $CO_2 \Rightarrow CO$ conversion efficiency, but most of the improvement occurs as the ratio increases from 1 to 2. For ratios above 2, further improvements are small. If one operated at a reactor pressure of say, 2.5 bars and used a H_2/CO_2 mixture ratio = 2, one could achieve CO_2 utilization efficiencies greater than 90%, provided that p (H_2O) $\sim$ 0.01 bar at 400 °C. This would require a hydrogen separation membrane and a hydrogen compressor.

Zubrin et al. (1997) operated a prototype RWGS reactor and they reported results roughly compatible with the curves in Fig. 1.14.

The flow chart for the RWGS reaction is shown in Fig. 1.15. It would appear that control of this process might require some meticulous care to be sure that the flow rates and heat transfer rates are all appropriate.

The RWGS system uses only one oxygen atom from each carbon dioxide molecule, and one molecule of CO is vented. Therefore, the requirement for pressurized CO_2 would be twice as high as for the S/E process at the same conversion efficiency. If we assume

Fig. 1.14 Calculated $CO_2 \Rightarrow CO$ conversion efficiency for the RWGS reaction for various H_2/CO_2 mixture ratios assuming p (H_2O) $\sim$ 0.01 bar at 400 °C

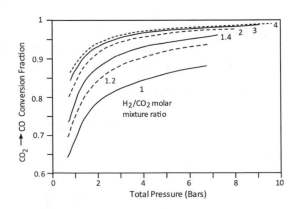

that the conversion efficiency of the S/E process is 0.95, and the CO$_2$ conversion efficiency of the RWGS process is w, then the requirement for pressurized CO$_2$ in the RWGS process is (2)(0.95)/(w) times that required for the S/E process. Note that if high reactor pressure is utilized to drive the RWGS reaction, this will put greater demands on the compressor that supplies pressurized CO$_2$.

The RWGS reactor can be used as a stand-alone process for producing oxygen for use with a fuel such as propane brought from Earth. In propane/oxygen combustion, oxygen accounts for 75–78% of the total propellant mass depending on the mixture ratio, and thus the RWGS system has considerable leverage even though it only produces oxygen. If the conversion efficiency can be made moderately high with hydrogen recovery (although this is made difficult by CO poisoning), the amount of hydrogen needed is quite small. The RWGS reactor can also be used in conjunction with a S/E reactor whereby the S/E reactor produces one molar unit of oxygen and one molar unit of methane, and the RWGS reactor produces one molar unit of oxygen, thereby attaining the proper stoichiometric

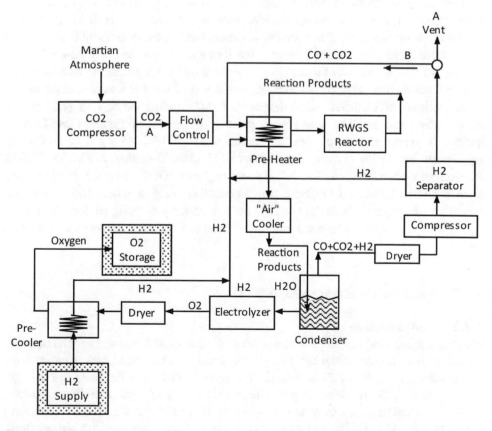

Fig. 1.15 Schematic flow diagram for reverse water gas shift process. Cryocoolers are not shown, but are attached to each of the cryo-tanks

ratio of roughly two molar units of oxygen per molar unit of methane. Zubrin has suggested combining the S/E process and the RWGS process in a single reactor, but this will be very difficult to make practical.

Zubrin et al. (1997) at Pioneer Astronautics researched the catalysts that have been used for the RWGS reaction, and they concluded that 10% copper supported on g-alumina has excellent properties. They operated a packed bed reactor using this catalyst and found that they produced pure CO with essentially zero side-products. The conversion efficiency without recirculation varied with flow rate, and reached 20% at the lowest flow rate, approaching the equilibrium conversion of 24% at 400 °C. They also operated the reactor with a condenser and recirculating compressor and obtained results compatible with the curves in Fig. 1.14, although a number of parameters are unclear in these results, particularly the ratio of recirculated flow rate to vented flow rate (B/A). Clearly, as B/A $\Rightarrow \infty$, one must approach the curves in Fig. 1.14, but the power requirements will also $\Rightarrow \infty$.

Holladay et al. (2007) demonstrated a microchannel apparatus for the RWGS process. It is not clear to this writer why anyone would want to use microchannel reactors for Mars ISRU processes, considering the high volumes of gases that must be throughput. Furthermore, despite the fact that almost any conceivable atmosphere acquisition system would have an elegant filtering system at the intake to reduce the likely possibility that some Mars dust might enter the reactor system, it would seem prudent to have sufficient reactor sizes so that minimal amounts of dust that do enter the reactor system can be accommodated. Nevertheless, Holladay et al. (2007) studied the RWGS process in a microchannel reactor. They were able to achieve conversion efficiencies close to equilibrium for reactor exit temperatures in the range 600–700 °C. As Fig. 1.12 shows, the equilibrium conversion at these temperatures is 35–45%. The selectivity for the RWGS reaction was about 98% at 700 °C. By running two RWGS reactors in series and removing water between the reactors, they increased conversion to over 50%. Comment: Incredibly, the paper by Holladay et al. did not refer to the pioneering work of Zubrin and co-workers who obtained higher conversion efficiencies at lower temperatures ten years earlier!

1.4.2 Solid Oxide Electrolysis (SOXE)

1.4.2.1 Introduction

Zirconia (ZrO_2) is a ceramic that can exist in several crystal arrangements. The cubic crystal is unstable because the Zr^{4+} ions are too small to fit the ideal lattice structure. This crystal structure can be stabilized by replacing up to about 10% of Zr^{4+} ions with a larger ion such as Y^{3+}. The resulting "doped" zirconia is called YSZ—yttria stabilized zirconia. Cerium or scandium can also be used as a dopant instead of Yttrium. The Y^{3+} ions form the equivalent of Y_2O_3 in the lattice; that is, each Y atom acquires 1.5 oxygen ions, whereas each Zr atom would like to acquire two oxygen ions. Therefore, the presence of

Y^{3+} in the ZrO_2 lattice takes oxygen ions away from the Zr^{4+} ions, producing oxygen vacancies within the lattice at Zr^{4+} sites.

As it turns out, these materials have an odd and unique property. When an electric field is impressed across a sheet of YSZ, oxygen ions can flow through the YSZ lattice, jumping from vacancy to vacancy until they arrive at the other side of the YSZ, whereupon they combine to form oxygen gas. Thus these materials can be electrically conductive, but the current is carried by O^{2-} ions rather than by electrons. This property offers the possibility of use as a fuel cell or an electrolysis cell, if porous, electrically conducting electrodes are attached, bounding both sides of the sheet YSZ. However, the conductivity is low at room temperature, and rises sharply with temperature. These systems must be operated in the 700–1000 °C range.

A CO_2 electrolysis cell is formed with a cathode layer, and electrolyte layer and an anode layer. The anode and cathode must be porous to allow gas flow, and electrically conducting. The cathode must contain material that acts as a catalyst for dissociation of CO_2. These layers must be made as thin as possible.

When heated CO_2 flows over the catalyzed cathode surface under an applied electric potential, a significant fraction of the CO_2 can be electrolyzed according to the reaction $CO_2 + 2e^- \Rightarrow CO + O^=$. The CO and any unreacted CO_2 is exhausted through an outlet tube, while the oxygen ions are electrochemically driven through the solid oxide electrolyte to the anode, where it is oxidized ($O_{2-} \Rightarrow O + 2e^-$). The O^{2-} ions combine at the anode to produce the gaseous O_2 that is released from the anode cavity at a rate proportional to current. The overall reaction is

$$[1]\, CO_2 \Rightarrow [X]\, CO_2 + [1 - X]\, CO + [(1-X)/2]\, O_2$$

Here, [1–X] is the fraction of CO_2 converted to CO, [X] is the fraction of CO_2 that is unreacted and flows out the exit tube, and [(1–X)/2] is the number of moles of O_2 produced per mole of CO_2 entering the cathode chamber. In typical applications, X is about 0.5.

The production rate of O_2 can be simply determined by measuring the ion current and dividing by (4 F), where F is Faraday's constant. In simple terms an oxygen flow rate of 1 g/h is equivalent to a current of 0.335 A.

In addition to the principal reaction of CO_2 electrolysis to produce oxygen, a variety of side reactions can also occur. These are generally detrimental to the cell and can lead to diminution of performance, and ultimately cell failure. The energetics of all relevant processes are understood, but the kinetics are less certain.

An objective of a solid electrolysis system is to maximize the throughput of oxygen production per unit area of the cells via the basic CO_2 electrolysis reaction. Faraday's Law allows us to calculate the reversible voltage requirement for the process:

$$CO_2(P_1) = CO(P_2) + 1/2\, O_2(P_3)$$

For $P_1 \sim 1$ bar and moderate conversion to O_2, the reversible voltage dependence on CO pressure is shown as the blue line in Fig. 1.16.

When the reversible voltage is applied to a cell, the tendency to act as an electrolysis cell ($CO_2 \Rightarrow CO + 1/2\ O_2$) is equally balanced against the tendency to act as a fuel cell ($CO_2 \Leftarrow CO + 1/2\ O_2$) and the slightest increase or decrease in the applied voltage will drive the reaction toward electrolysis or a fuel cell. This reversible voltage is also referred to, as the "open circuit voltage" because a cell left unconnected will develop that voltage as if it were a battery. To operate as an electrolysis cell, a voltage greater than the open circuit voltage must be applied in a polarity opposite to the open circuit voltage. The effective net voltage producing electrolysis is $(V - V_o)$ where V is the externally applied voltage and V_o is the open circuit voltage.

It is desirable to obtain a high value of the current density (A/cm^2). This allows generation of a high flux of oxygen for any given cell area. The current through the cell is (i) amps which is equal to

$(V - Vo)/R$, where R is the cell resistance. The current density is ($I = i/A$) where A is the cell area. Thus

$$I = i/A = (V-V_o)/(R\,A)$$

The efficiency of a cell is inversely proportional to the cell resistance (R). But the cell resistance is inversely proportional to the cell area. Hence the product (R A) provides an area-independent measure of the innate cell resistance, which is an inverse measure of cell efficiency. Thus we define the area-specific resistance (ASR) of a cell as

$$ASR = (R\,A) = (V-Vo)/I$$

The ASR provides an inverse measure of cell efficiency. Lower ASR results in higher throughput. It has been found by experiment that the simple picture of a cell described

Fig. 1.16 Dependence of reversible voltage requirement versus CO pressure. The red curve is for reduction of CO to C, and the blue curve is for electrolysis of CO_2

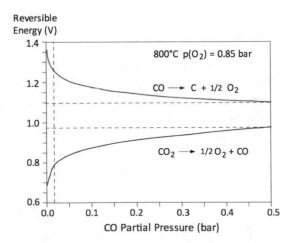

above is not accurate. If one plots current vs. voltage, the current intercept at zero voltage lies higher than the theoretical open circuit voltage. Thus the net effective voltage for oxygen formation is V(net) = {cell voltage − OCV − V(act)} where V(act) is an empirical activation voltage, and OCV is the open circuit voltage. The slope of the I–V curve determines the true intrinsic ASR(i). Then the cell current is V(net)/ASR(i).

The minimum electric energy supply required for the electrolysis process is equal to the change in the Gibbs free energy:

$$\Delta G = \Delta H - T\Delta S$$

where ΔH is the enthalpy change, T is the temperature and ΔS is the entropy change.

The SOXE can theoretically work at the so-called thermo-neutral condition. At the thermo-neutral voltage (V_{tn}), the electrical power input exactly matches the total power demand of the electrolysis reaction. In this case, the electrical-to-oxygen conversion efficiency is 100%. At cell operating voltages < V_{tn}, heat must be supplied to the system to maintain the temperature and the conversion efficiency (based only on the electrical input) is above 100%. At cell operating voltages > V_{tn}, excess heat must be removed from the system and the efficiency is below 100%.

At the typical temperature of operation (~ 800 °C), V_{tn} is around 1.46 V. At this level, the cell can theoretically be operated at thermal equilibrium with an electrical utilization efficiency of 100%. However, there are a number of reasons to operate below the thermo-neutral voltage to reduce the incidence of side-reactions.

1.4.2.2   Background

A solid oxide device can be operated as a fuel cell or as an electrolysis cell, with opposite voltage polarity. A considerable amount of work has been done on solid oxide fuel cells, with less work done on electrolysis, particularly electrolysis of CO₂ without water present. Even less has been done on electrolysis of pure CO₂.

Much of the early work on solid oxide electrolysis of CO₂ utilized porous Pt electrodes in single wafer systems. Platinum is a good catalyst for this purpose. However, it is difficult to control the layer thickness and uniformity of the electrodes. The key to making solid oxide electrolysis practical is to develop a system of electrodes and electrolyte that utilizes thin, uniform, flat, reproducible elements, so that multiple cells can be stacked, one upon another, to allow a high throughput within a small enclosure maintained at high temperature. It is not clear whether this can be done with platinum electrodes.

For many years, most experimental work was confined to use of YSZ tubes coated with porous platinum electrodes (e.g. Richter 1981; Ramohalli 1991) but most later work was done on more compact and potentially rugged flat disk geometries.

Several approaches to utilizing YSZ flat disk stacks were reported in the 1990s. One of these was led by Crow and Ramohalli (1996) of the University of Arizona and was identified by the name "MOXCE" (Crow 1997). A central innovation of MOXCE was that the ion conductors were disks cut from single crystals of YSZ, available commercially as substrates for the microelectronics industry. A second innovation was that the other cell components were made of an alloy of platinum with 10–20% rhodium. Platinum has the disadvantages of being heavy and costly, but it has two advantages that make it the material of choice: it does not react with YSZ, and it has a coefficient of thermal expansion almost identical to that of YSZ. Both have coefficients of thermal expansion of $1.0 \times 10^{-5}/°C$ and are identical within the limits of engineering accuracy.

MiniMOX was a small device operated at a pressure of 1 atm. The lower cell cap was grounded and served as the cathode, while the upper cell cap was the anode and had a positive voltage. The disk was a single crystal made of 9.5 mol percent yittria stabilized zirconia. The diameter of the disk was 5.0 cm and nominal area was 19.64 cm^2. The cell walls were annular rings that pressed against the outer 4 mm of disk radius, leaving 13.85 cm^2 exposed to the gases. No attempt was made to seal the edges in this simple proof-of-concept experiment, so air leakage around the edges had to be accounted for. An electrical phenomenon further reduced the effective area of the disk. The electrodes started as platinum ink, were silk screened onto the disk faces, then fired, and finally "broken in" by running MiniMOX for 20 or more hours to make them porous. The electrodes were about 10 μm thick, but porosity made their radial conductivity somewhat conjectural. Whatever may have been the exact conductivity, the voltage drop from the perimeters of the electrodes to their centers was clearly significant. To the extent that the inner portion of a disk did not "see" full voltage, it would have been less effective in supporting electrolysis. Therefore the effective area was probably less than 13.85 cm^2. The thickness of a disk was 0.5 mm.

MiniMOX was subjected to eight thermal cycles from 740 to 1000 °C and 42 voltage cycles from 0 to 2 V with no ill effects. The cell disassembled easily after the tests, and the YSZ disk was intact.

Because the outer edge of the YSZ disk was merely press-fit against the platinum supports with no sealant applied, some air leaked in at the edges and provided an oxygen supply at the cathode to be pumped to the anode. Therefore, the apparent ion currents measured were greater than that which would have been obtained if 100% of the CO$_2$ had been converted to O$_2$. An attempt was made to correct for this effect by running the cell with argon flowing though it, and assuming that the observed ion current represented air leakage at the edges. Subtracting this current off from that observed with CO$_2$ led to the final results that are represented approximately in Fig. 1.17. Crow and Ramohalli claimed that they could approach 100% conversion of CO$_2$ to O$_2$ in their cell at the highest temperatures and voltages. It is not clear from their report whether an carbon formation took place. Except for minor anomalies, the MiniMOX results appear to indicate that cells using single crystal YSZ are thermally robust and perform close to the level predicted by theory. There was an indication that the conversion efficiency could be run up close to 100% but the disturbing edge effects introduce uncertainty.

MOXCE 2 was designed as a two-cell system, scalable to the production rates of the foregoing section. The disks of MOXCE 2 were 10.16 cm in diameter, about twice the diameter of the MiniMOX disk, so MOXCE 2 should have produced oxygen at about eight times the rate of MiniMOX. The cells of MOXCE 2 were wired in series, so the voltage added up with number of cells, but the current was fixed at about four times the current of MiniMOX. It was heated and cooled several times and seemed to show no adverse effects. Then a stack was subjected to an impressed voltage to activate the electrodes but it did not perform properly. When the stack was taken apart, it was found that one disk had cracked and the front electrode had delaminated. While there are a few theories of what might have happened, it is not certain what caused these problems. It seems possible that the two wafers "burned in" at different rates and the voltage drop across one wafer was much larger than the other causing a failure of that wafer. This work was not completed for lack of NASA funds.

Additional difficulties can be encountered in YSZ stacks that are not encountered in single wafer units. It is not clear that one can achieve a good balance between resistances of individual wafers, and it may be necessary to energize each wafer in parallel if a series circuit produces imbalances in voltage drops across each wafer.

The greatest challenges in these devices are to:

(1)   Make them physically rugged and able to withstand shaking and vibration.
(2)   Provide a large surface area of YSZ in a small volume. This, in turn requires an intricate manifold/plenum system that carries CO$_2$ to one edge of each disk, spent CO + CO$_2$ away from the other edge of each disk, and pure oxygen to a separate reservoir from the reverse side of each disk.
(3)   Make them resistant to thermal shock due to heat-up/cool-down cycles, and particularly heat-up cycles that must raise the temperature to working temperature in ~60–90 min.
(4)   Apply electrodes that remain integral with the electrolyte and function through many cycles without delaminating.
(5)   Seal all edges and joints, while allowing for thermal expansion and contraction and connecting the ceramic YSZ to metallic exit tubes.
(6)   Achieve a reasonable balance of wafer resistance from wafer to wafer.

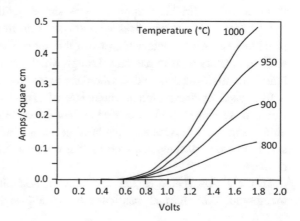

Fig. 1.17 Ion current obtained by Crow and Ramohalli using "Minimox" at 60 sccm CO$_2$ flow rate

(7) Utilize as high a fraction of CO_2 as possible in order to minimize the required capacity of the CO_2 compressor that supplies high pressure CO_2 to the zirconia cell.

K. R. Sridhar had a continuing program in YSZ cell development at the University of Arizona in the 1990s—assisted by support from the Hamilton-Standard Corporation, who was interested in miniature oxygen pumps to separate oxygen from air and compress it for medical and aeronautical applications on Earth. Although this application is rather different than ISRU on Mars, the hardware has many things in common. Sridhar also worked on adapting his oxygen pump hardware to processing CO_2 on Mars. Sridhar utilized ceramic disks with glassy materials for sealing the edges.

Sridhar built a number of YSZ cells using tubular and flat disk geometries. Sridhar and Miller (1994) operated a single disk cell (see Fig. 1.18). It is not clear how the elements of this cell were sealed, nor is it clear how ceramic/metal seals were made to the three ceramic exit tubes. They obtained current densities in the range 0.1–0.35 A/cm^2, depending on the temperature and applied voltage. No mention was made of the ability of this cell to withstand thermal cycling. The curves of current density versus voltage were similar in general form to those shown in Fig. 1.17, with highest current density being just under 0.4 A/cm^2, achieved at 1000 °C at 1.9 V.

Sridhar and Vaniman (1995) reported further tests with the device shown in Fig. 1.18. They verified that indeed the observed current was equal to the oxygen flow rate, via the relationship 1 A = 0.335 g/h of O_2, thus demonstrating that no spurious currents were involved. They also claimed that such a cell can endure for 1000–2000 h without noticeable degradation in performance. However, they did not seem to mention whether this cell could endure thermal cycling. Sridhar and Vaniman (1995) found that when the ion current of the cell was plotted versus CO_2 flow rate, the ion current was independent of flow rate at sufficiently high flow rates, and diminished when the flow rate was reduced below a certain point. In general, this led to CO_2 conversion efficiencies in the 30–40% range.

Sridhar also claimed that he developed ceramic YSZ cells for stacking into multiple wafer cells, although there doesn't seem to be any published data on these devices (Sridhar 1995). Sridhar's approach was similar to that used by Suitor et al. (1990). It involved casting ceramic disks with structure on their surfaces to allow gas flow while providing electrical contact via ribs to the entire surface of the YSZ. These disks were stacked together to form plenums through which gas flows radially through each disk, and longitudinally from disk to disk where the holes line up. The whole conglomerate is sealed with a glass having a similar coefficient of expansion curve to that of YSZ over wide temperature ranges. Sridhar carried out considerable modeling to assure that flow will be uniform in these devices and that heating will be uniform. Unfortunately, there does not seem to be any data available on the performance and ability to thermally cycle of such devices.

In the 1990s, Allied Signal developed a YSZ alternative using tape-calendered thin membrane YSZ formed continuously as a 3-layer sandwich with electrodes. The

Fig. 1.18 One-disk YSZ
device (Sridhar and Miller
1994)

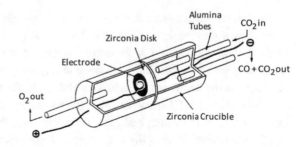

fabrication method developed by the Allied Signal Company for YSZ cells was based a
tape calendering process developed for producing solid oxide fuel cells (SOFCs) with thin
(1–10 μm) YSZ electrolytes. For SOFCs, the use of thin YSZ electrolytes reduces ohmic
losses.

The tape calendering process for making thin electrolyte cells involves progressive
rolling or calendering of green (unfired) tapes to produce a thin electrolyte on a support
electrode. In this process, electrolyte (YSZ) and the anode layer powders are first mixed
with organic binders and plasticizers in a high-intensity mixer to form a plastic-like mass.
The mass is then rolled into tapes using a two-roll mill. YSZ and anode tapes of specific
thickness ratios are laminated and rolled into a thin bi-layer tape. This thin bi-layer tape is
then laminated with an anode, and the laminate is rolled again into a thin bi-layer tape.
This progressive rolling process can be repeated with different tape thickness ratios until a
desired YSZ film thickness is obtained. In general, the process requires only three rollings
to achieve a bi-layer with a YSZ film less than 10 μm thick. The final bi-layer is fired at
elevated temperatures to remove the organics and sinter the ceramics. A cathode layer is
applied on the electrolyte surface of the sintered bi-layer to produce a complete cell. It
should be noted that the cathode can be rolled along with the electrolyte and anode to form
a tri-layer. In this case, a separate cathode application step is not needed, and the tri-layer
can be co-fired in a single firing step. Initial testing used the bi-layer approach to allow use
of a variety of cathode materials with a fixed anode configuration.

The calendering process was adjusted to create a dense YSZ central portion of the triple
layer sandwich, with porous electrodes on either side of the YSZ. Allied Signal fabricated
a number of single wafer cells and tested them with CO_2. A variety of electrolyte
thicknesses were tried, with several combinations of materials for the cathode and anode.
A platinum grid was included in the electrode to assure uniform electrical potential across
the wafer, and this also provided a simple means of electrically connecting to the wafer.
Based on the curves of I versus V, they tentatively selected a combination involving a
13-μm thick YSZ electrolyte, with a platinum/YSZ composite cathode, and an LSM/YSZ
composite anode. This wafer yielded a current density of about 0.2 A/cm^2 at 1.4 V, and
about 0.5 A/cm^2 at 1.7 V, at 900 °C. They thermally cycled this wafer between 30 and
900 °C 10 times and re-measured the I versus V curve and found it to be essentially
identical.

In general, one expects that the I versus V curves for a YSZ disk will have the rough form shown previously in Fig. 1.17. There will be essentially no ion conductivity below about 1 V, and above ~ 1 V the current density will be almost proportional to the voltage, indicating a roughly constant resistance at any temperature. The upward slope of a line in Fig. 1.17 is the inverse of the resistance of the wafer at that temperature. The resistance is determined by $R = KL/A$, where K is the resistivity (dependent on temperature only), L is the thickness of the wafer, and A is the face area of the wafer. The object is to make R as small as possible. Raising the temperature decreases K; however it is desirable to operate at as low a temperature as possible to minimize heat losses and thermal stresses during cycling. For fixed A, R is determined by the product KL. If L can be made thin enough, one can operate at lower temperatures and still attain acceptable values of the product KL. Allied Signal believed that they could achieve thinner zirconia cells with their method, and thereby operate at lower temperatures than others in this field.

Allied Signal assembled these wafers into 3-wafer stacks using the approach adapted from Allied Signal SOFCs. This design incorporated zirconia cells in a compliant metallic structure of cross-flow configuration. The cells were connected in electrical series using metallic fins and interconnects. The fins attached to the interconnects (forming interconnect assemblies) formed flow channels for reactant and product gases. The interconnect assembly was designed to provide compliancy to minimize the stresses due to thermal expansion mismatch and form a compact, lightweight structure. In this approach, the stack was made up of modules with each module consisting of (i) a carbon dioxide manifold to deliver CO_2 to cathodes above and below it, (a mixture of $CO + CO_2$ exits on the other side of the manifold), (ii) a 3-layer wafer consisting of a cathode, electrolyte and anode, (iii) each of the anodes on the other side of each 3-layer wafer rests against an oxygen manifold to carry oxygen out of the cell, and (iv) each oxygen manifold mates to a platinum interconnect sheet. These modules can be stacked, one on top of another. All of the oxygen manifolds were connected to an oxygen plenum, all of the CO_2 manifolds connect to the CO_2 plenum, and all $CO + CO_2$ manifolds were connected to the exhaust plenum.

Initial tests were only mildly encouraging. Difficulties were encountered in sealing the manifolds. Imbalances in resistances of three wafers connected in series prevented the stacks of three from being fully activated. Nevertheless the stack of three wafers did survive several thermal cycles. It is unclear how much additional effort and funding would be required to make this technology viable, or even if it could have been made viable at all. Lack of NASA funding left this program unfinished.

In the late 1990s, NASA built the Mars ISRU Precursor (MIP) that was scheduled to fly on a Mars lander mission that was never funded and built. The MIP included a single cell solid oxide electrolysis device that was tested in the laboratory (Fig 1.19).

It is noteworthy that Fig. 1.16 shows that at voltages above about 1.1 V, carbon formation via dissociation of CO becomes thermodynamically possible. Yet, none of the studies done in the 1990s with Pt electrodes reported carbon formation. It is not clear whether the kinetics of carbon formation are inhibited in the presence of Pt, or perhaps

Fig. 1.19 Performance curves for single cell MIP SOXE

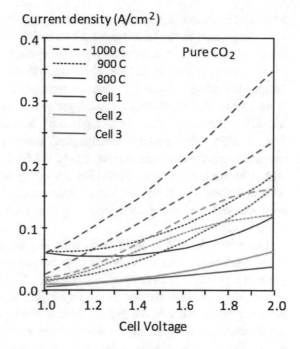

whether there were voltage drops across the electrodes, so that the actual voltages across the electrolyte were lower.

A considerable amount of work was done in the early 21st century aimed at applications to solid electrolyte fuel cells. In this work, new physical structures and new catalytic coatings for YSZ were developed that appear promising. However, most of this work does not appear to provide data directly appropriate to electrolysis of CO_2 to form O_2 under Martian conditions. A few relevant papers include: Kim-Lohsoontorn and Bae (2011), Ebbesen and Mogensen (2009) and Bidrawn et al. (2008). All of this work involved single disks of solid electrolytes.

Tao et al. (2004) provided a comparison of performance of CO_2 electrolysis using Pt electrodes and Pt-YSZ cermet electrodes. The cells with Pt-YSZ cermet electrodes had advantages at low current densities but these diminished at higher current densities.

Ceramatec, Inc. developed a system for stacking large numbers of planar cells for fuel cells.

1.4.2.3 MOXIE

In 2014, NASA selected a team led by MIT and JPL to implement a demonstration of solid oxide electrolysis of CO_2 to produce oxygen on Mars. The resultant Mars Oxygen ISRU Experiment (MOXIE) Project selected Ceramatec, Inc. as its contractor to develop the SOXE subsystem, based on past success in demonstrating stacks of solid oxide fuel cells. Ceramatec has traditionally produced externally manifolded stacks for terrestrial operations in order to maximize the fraction of electrolyte active area as a means of scaling stacks to the largest practical footprint within the limitations of producing thin flat ceramic cells at an acceptable yield. However, the MOXIE mass, volume and sealing requirements all favored an internally manifolded stack design. For MOXIE, Ceramatec designed a highly compact, structurally rigid stack using internal manifolding and glass seals. Ceramatec and Plansee worked together to develop a custom interconnect design for MOXIE, addressing both functional and fabrication considerations, based on the Plansee CFY alloy powder metallurgical net-shape process. The CFY interconnect material thermal expansion matches that of the ESC so closely that the two can be joined with a hermetic glass seal and reliably cycled. In contrast, the thermal expansion match with low chromium ferritic stainless interconnects developed at Ceramatec, while good, only allowed the use of compressive gasket seals effective only with small pressure differentials.

The remainder of the materials set is largely the same as developed by Ceramatec for steam electrolysis. The electrolyte is tape cast scandia stabilized zirconia. The anode is a proprietary perovskite developed by Ceramatec. The cathode is an oxide dispersion nickel cermet. Consideration was given to using copper based electrodes, which are reported to inhibit carbon deposition. Replacing nickel with copper in the cathode does not change the equilibrium thermodynamics of carbon deposition in the CO/CO_2 system. While nickel is more effective at templating carbon nucleation than copper, statistical mechanics suggest that nucleation is still possible on any surface given sufficient time or area exposed to gas compositions that thermodynamically favor carbon deposition. The electrochemical kinetics of the nickel cermet toward CO_2 reduction are superior to the copper based electrode formulations known to Ceramatec. Given the importance of this mission, it was determined to rely on a carbon avoidance strategy based on equilibrium thermodynamics bounded gas compositions rather than relying on statistical mechanics limited nucleation kinetics with gas compositions favoring carbon.

MOXIE was required to demonstrate delivery of high purity oxygen, at a nominal pressure of 1 bar while operating in a nominal 10 mb environment. This suggested a 3-port stack design, the cathode flow fields having a CO_2 inlet and CO/CO_2 outlet and the anode flow fields having a single O_2 outlet port, all internally manifolded. The stack mass and energy allocations dictated the use of a pair of five cell stacks in series, within a volume envelope of 50 mm $\times$ 100 mm $\times$ $\sim$30 mm. An intermediate busbar on the interconnect between cells 5 and 6 allows independently powering cells 1–5 and 6–10 as sub-stacks should one sub-stack show high resistance after landing. However the entire stack of 10 cells has common CO_2 feed and O_2 collection flow paths, so no redundancy against gas pressure containment or flow obstruction is available. The MOXIE stack configuration is shown in Figs. 1.20 and 1.21.

Early tests directed toward Mars/MOXIE operating conditions were conducted on a number of preliminary test stacks. After testing various stacks, refinements were made on the design of subsequent test stacks. Comparison was made between use of pure CO_2 and *Mars mixture gas* (MMG) and the effect of trace quantities of CO and O_2 was studied.

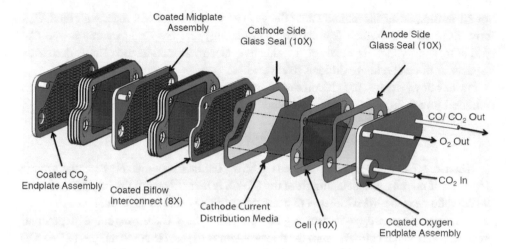

Coated Midplate Assembly
Cathode Side Glass Seal (10X)
Anode Side Glass Seal (10X)
CO/ CO_2 Out
O_2 Out
CO_2 In
Coated CO_2 Endplate Assembly
Coated Biflow Interconnect (8X)
Cathode Current Distribution Media
Cell (10X)
Coated Oxygen Endplate Assembly

Fig. 1.20 Exploded view of MOXIE stack

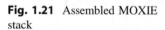

Fig. 1.21 Assembled MOXIE stack

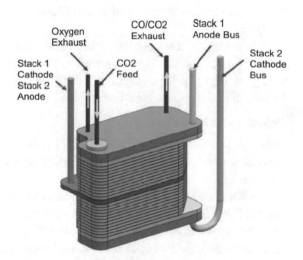

Oxygen Exhaust
CO/CO2 Exhaust
Stack 1 Anode Bus
Stack 2 Cathode Bus
Stack 1 Cathode Stack 2 Anode
CO2 Feed

Startup/shutdown thermal cycles were carried out and ambient leak tests were conducted between cycles. Startup from ambient to 800 °C typically had a duration of 90 min. Longer duration operation was also studied to some extent, and performance degradation was monitored. Cell-to-cell performance (voltage), stack current, O_2 production, and CO_2 conversion were monitored.

Testing was carried out on a series of test stacks. The following issues arose and were dealt with in subsequent work.

Oxidation: No voltage

If a hot (~ 800 °C) cathode is exposed to CO_2 with no voltage applied, significant structural damage to the cathode can result. As CO_2 (in the absence of CO) contacts the hot Ni in the part of the cathode near the gas entry, the nickel will react with the CO_2 to form CO and NiO (which is electrically insulating) and the Ceria returns to the Ce^{4+} oxidation state which is also not an effective electronic conductor. These destructive impacts to the cathode are difficult to reverse.

Further downstream, the CO produced by reduction of CO_2 upstream will protect the cathode by reducing NiO via the reaction

$$NiO + CO \Rightarrow Ni + CO_2$$

When the CO concentration exceeds 0.55%, oxidation of the Ni essentially stops. However, if no tail gas emanating from the SOXE is recirculated to the inlet, the region of the cathode near the MMG inlet is exposed to pure CO_2, and as it gradually degenerates, CO will not be produced in that region; therefore the CO-free region will gradually propagate across the cell. The cure for this problem is to recycle some tail gas (CO + CO_2 effluent from SOXE) to the inlet of the scroll pump so the CO concentration is always >0.55%, even at the inlet port to the SOXE. After further study, the MOXIE Project included recirculation of some tail gas to the entry so there is always CO present at >0.55%.

It is imperative that the hot cathode must not be exposed to CO_2 when there is no electric power applied to the electrodes. The sequence of operation must be such that voltage is applied prior to admitting MMG to the SOXE. In the advent of a sudden power failure, the scroll pump must be turned off quickly.

Oxidation with voltage applied

Although exposure of the cathode to MMG is far less deleterious when power is on than it is when power is off, nevertheless oxidation of Ni in the cathode can still occur when power is on.

If a hot (~ 800 °C) cathode is exposed to MMG when voltage is applied, the applied potential can theoretically electrochemically reduce any previously built-up NiO to Ni, with the oxygen in NiO being converted to O^{2-} ions. In practice this doesn't work so well since the conductivity of oxidized cathode is poor. Nevertheless, it can have some effect if the cathode isn't too far oxidized. In addition, the reaction of electrochemically produced CO (from CO2) can reduce NiO via:

$$NiO + CO \Rightarrow Ni + CO_2$$

This process is effective when the CO concentration is >0.55%.

Another mechanism that helps protect the cathode during electrolysis operation is the rib pattern. There are long channels with primary flow, and there are breaks between these channels every few mm for the purpose of flow distribution equalization. For the most part there is little cross channel flow, but there is CO production in these areas, and also under the ribs (at least near the perimeter of the rib footprint). These areas are harder to oxidize with flowing CO_2 because they are "sheltered from the wind" so to speak, but they still generate CO that flows into the main channels.

Recirculation of cathode tail gas

Recirculation of cathode tail gas (containing CO) to the CO_2 incoming to the SOXE, serves to protect the MMG inlet region from oxidation, and even reverse light oxidation near the inlet by NiO reduction in reaction with the recycled CO. If the inlet gas to the SOXE contains 2% CO, oxidation should be essentially eliminated. If some of the cathode tail gas (CO + CO_2) is recirculated back to the entry of the scroll pump, the gas entering the SOXE will contain some CO along with the major portion of CO_2.

A combination of recirculation of tail gas, plus controlled start-up and shutdown procedures will essentially prevent oxidation of the Ni catalyst.

Carbon formation by reduction:

If carbon formation were to occur, it could severely impact cell performance. Solid carbon can destroy electrode microstructure and obstruct gas flow channels in the stack. Furthermore, remote ISRU applications offer little opportunity to recover from a carbon deposition episode by conducting an oxidation and reduction cycle on the anode.

Carbon formation can occur via three processes

(1) $2CO \Rightarrow C + CO_2$ (Boudouard reaction)
    The CO is formed by

$$CO_2 + 2e^- \Rightarrow CO + O^{2-} \text{(electrode reaction)}$$

(2)   CO electrolysis: $CO + 2e^- \Rightarrow C + O^{2-}$ (electrode reaction)
    The CO is formed by

$$CO_2 + 2e^- \Rightarrow CO + O^{2-} \text{(electrode reaction)}$$

(3)   Direct CO_2 electrolysis to carbon

$$CO_2 + 4e^- \Rightarrow C + 2O^{2-} \text{(electrode reaction)}$$

The net reaction in all three cases is the same as for reaction (1.3).
The chemical process is called the Boudouard reaction:

$$2CO \Rightarrow C(s) + CO_2$$

At 800 °C and 1 bar, CO will disproportionate and deposit carbon only when the CO concentration exceeds about 89%. Since the SOXE in MOXIE is always operated at lower utilization (typically up to 50%) it would appear at first glance that there is no problem here—and indeed there may well be if the cell is optimally designed. However, the distribution of CO_2 and CO in the flow channels through a cell is not uniform. The three-phase boundary where CO_2 diffuses into a region of the cathode where the Ni catalyst is present and electric voltage is reached is where the electrolysis reaction takes place. This requires CO_2 to diffuse from the flow channel, through the pores of the cathode to the three-phase boundary region, react, and then CO must diffuse back to the flow channel. This requires a concentration gradient of CO_2 from flow channel to three-phase boundary region, and a concentration gradient of CO from three-phase boundary region to the flow channel. All this occurs without any pressure gradient driving flow in the transport direction. Thus the CO partial pressure at the three-phase boundary must be higher and the CO_2 partial pressure must be lower than in the free stream channel flow. It isn't the bulk exit average CO concentration in the tail gas that matters here; it is the worst-case (highest) CO deep in the pores of the cathode. If these effectively exceed the 89% threshold, coking can occur. At 850 °C the threshold rises to 97%. However, higher operating temperatures leads to extra stress on the seals. Lower voltages produce less coking, but there is less oxygen production. Lower operating pressures also help but that is not feasible with the current implementation of MOXIE.

Carbon can also form electrochemically via reaction (2). That is why MOXIE is operating the SOXE at lower voltages than might otherwise be utilized. We already showed the voltage required to electrolyze CO to solid carbon (red curve in Fig. 1.16). This curve shows that the threshold for electrochemical carbon formation varies along the cell from over 1.2 V at the gas entry to as low as 1.1 V at the gas exit if there is 50% conversion. In order to assure that electrochemical carbon formation does not occur, the voltage at the gas-cathode-electrolyte interface must be maintained below about 1.1 V. If the voltage impressed upon each cell is 1.1 V (or less), electrochemical carbon formation should be eliminated. However, as we showed previously, the current density is

$$I = i/A = (V-Vo)/(R\,A)$$

Operating at V $\sim$ 1.1 V (or lower) limits the current density. It would be desirable to operate at higher voltages if carbon formation could be avoided.

Inconsistent Cell Resistance

The cells in a stack are wired in series. The current is the same through all the cells. Therefore the voltage drop across each cell is proportional to the cell resistance.

Figure 1.16 shows curves of reversible energy for (1) reduction of CO_2 to CO + O_2; and (2) for reduction of CO to C and O_2. In order for oxygen production to occur, the voltage on a cell must exceed the lower curve at any CO partial pressure. But the voltage must be below the upper curve to assure prevention of electrochemical carbon formation. At any CO_2 flow rate, the voltage is set to maximize the oxygen production rate, while keeping the voltage below that which would cause carbon formation. Here, there are two opposing considerations: the higher the voltage, the greater is the oxygen production, but the sooner does the operating point intersect the Nernst curve for carbon formation. The maximum oxygen production rate is not necessarily achieved by maximizing the percent conversion. At high flow rates it is advantageous to use a higher voltage with a moderate percent conversion. At low flow rates it is advantageous to use a lower voltage and obtain a higher conversion rate. This is illustrated in Fig. 1.22. The upper solid horizontal lines show applied voltages for three CO_2 flow rates: 30, 60 and 100 g/h. The lower dashed curves show the average of the CO_2 Nernst potential over the range of percent conversion for that flow rate. The vertical red arrows show the average "driving force" for oxygen production at each flow rate.

By judiciously choosing voltages, the percent utilization and oxygen production rate at each flow rate are obtained as shown in Fig. 1.23. Note that the oxygen production rate at 30 g/h is about half of the oxygen production rate at 100 g/h, even though the flow rate is only 30%. Figure 1.23 shows ideal theoretical curves, assuming all cells have the same ASR, so that the voltage applied to the stack will tap off equally among the cells, and that voltage can be taken almost up to the Nernst voltage for carbon formation on each cell, thus maximizing oxygen production while avoiding carbon formation. In actuality, each cell will have a slightly different ASR. The cell with the highest ASR will have the greatest voltage drop across it. Normally one might apply say, 10.8 V across a 10-cell stack and hope to drop 1.08 V across each cell at 60 g/h flow rate. That would be the ideal. But if one cell in the stack had a higher ASR so its voltage drop would be say, 1.11 V that might allow carbon formation on that cell. Hence, the voltage applied to the whole stack would have to be cut back to say, 10.5 V in order to keep the voltage drop across the most resistant cell down to 1.08 V. But now, most of the cells would be operating at 1.05 V, and oxygen production would be lower than the ideal. Thus the red curve in Fig. 1.23 provides an upper limit to the oxygen production rate, and test data on cells lie somewhat below this red curve. Similarly the blue curve in Fig. 1.23 provides an upper limit to the percent utilization, and test data on cells lie somewhat below the blue curve.

By judiciously choosing voltages, the percent utilization and oxygen production rate at each flow rate are obtained as shown in Fig. 1.23. Note that the oxygen production rate at 30 g/h is about half of the oxygen production rate at 100 g/h, even though the flow rate is only 30%. Figure 1.23 shows ideal theoretical curves, assuming all cells have the same ASR, so that the voltage applied to the stack will tap off equally among the cells, and that voltage can be taken almost up to the Nernst voltage for carbon formation on each cell, thus maximizing oxygen production while avoiding carbon formation. In actuality, each cell will have a slightly different ASR. The cell with the highest ASR will have the greatest voltage drop across it. Normally one might apply say, 10.8 V across a 10-cell stack and hope to drop 1.08 V across each cell at 60 g/h flow rate. That would be the ideal. But if one cell in the stack had a higher ASR so its voltage drop would be say, 1.11 V that might allow carbon formation on that cell. Hence, the voltage applied to the whole stack would have to be cut back to say, 10.5 V in order to keep the voltage drop across the most resistant cell down to 1.08 V. But now, most of the cells would be operating at 1.05 V, and oxygen production would be lower than the ideal. Thus the red curve in Fig. 1.23 provides an upper limit to the oxygen production rate, and test data on

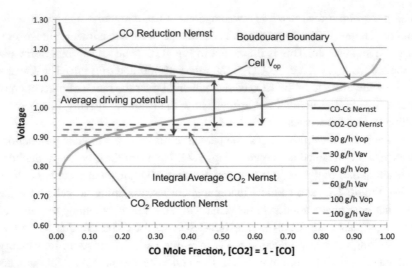

Fig. 1.22 SOXE operational envelope

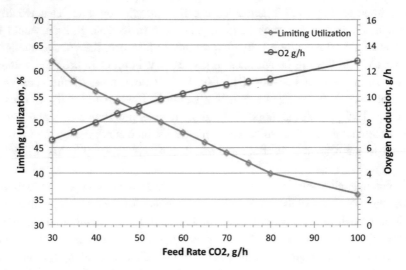

Fig. 1.23 SOXE operational envelope

cells lie somewhat below this red curve. Similarly the blue curve in Fig. 1.23 provides an upper limit to the percent utilization, and test data on cells lie somewhat below this blue curve.

1.4.2.4 Operating Pressure and Temperature

The obvious convenient choice for operating pressure is about one bar, which is compatible with testing on Earth, and allows good gas throughput in small tubes.

The choice of operating pressure for MOXIE is ultimately a system issue, more than a SOXE issue.

The difficulty in designing a compressor increases significantly as its compression ratio increases. With SOXE operating at 1 bar, the required compression ratio is in the range 150–250 depending on the location and season on Mars. As we discussed previously, the true compression ratio of the present scroll pump is about 6:1, but the much higher effective compression ratio is achieved by repeated positive displacements of gas into the chamber downstream of the pump. Yet, because the pressure in that chamber exceeds the pressure in the pump, fluctuations in pressure occur during the pumping cycle. At lower overall compression ratios, this becomes less problematic. Lowering the overall compression ratio is bound to provide significant benefit to the scroll pump.

The safe operating region (as shown for example in Fig. 1.22) increases at lower operating pressures (see Fig. 1.24). This would provide an additional incentive for lowering the cathode pressure for MOXIE operation, since it would allow higher voltages without coking. It has been estimated that at 0.5 bar, the allowable higher voltage could increase the current density by about 6%.

The operating temperature also has implications for allowable voltage. At higher temperatures, the $CO \Rightarrow C$ curve in Fig. 1.24 moves up, the cell resistance is reduced, and the CO_2 reduction voltage is reduced.

However at higher temperatures, the preheat energy increases, thermal losses increase, seals become more difficult, and the solid state reaction rates at the core of most other degradation mechanisms might also increase.

1.4.2.5 Oxygen Production Rate

Viking measurements showed that surface temperatures at an equatorial location can vary from 280 K at solar noon to 170 K before dawn (Wilson and Richardson, 2000).Viking measurements (Wood and Paige 1992) also indicate that local pressure variations during the course of a Martian year can fluctuate by about ±15% about the mean. Thus, for any equatorial landing site, we can expect variations of about ±15% about the mean pressure, depending on season. The diurnal variation in atmospheric temperature is wide.

The density of Mars atmosphere varies as shown in Fig. 1.25.

A likely landing site for the 2020 Mars Rover is Jezero Crater, where the nominal pressure averaged over a Martian year is 5.25 Torr. During the course of a Martian year, the actual pressure will vary by ±15%, or from about 4.5 Torr to 6.0 Torr. The daily diurnal variation in temperature is estimated to be from 170 K to 280 K from before dawn to solar noon. According to Fig. 1.25, the density of incoming Martian atmosphere can vary from about 11.5 g/m³ at 280 K and 4.5 Torr, to 24.5 g/m³ at 170 K and 6 Torr (more than a factor of two). However, if the entering gas is warmed in the inlet tube, the higher density might not be achieved.

Since the scroll pump is a volumentric delivery device, the mass flow into the ISRU system is proportional to the Mars gas density. Based on empirical data as well as

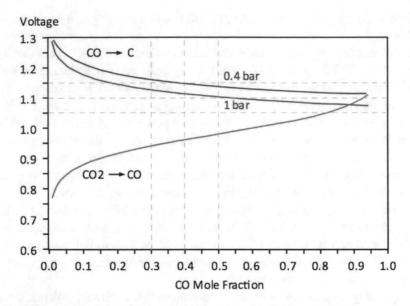

Fig. 1.24 Effect of lowering cathode pressure from 1 bar to 0.4 bar at 800 °C

Fig. 1.25 Density of Mars atmosphere as a function of T and p

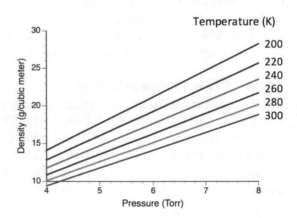

Table 1.10 Modeled operating points

CO$_2$ flow rate (g/h)	CO$_2$ utilization rate (%)	Current (A)	Voltage	g/h of O$_2$ produced per cell	Apparent ASR
30	62	2.15	1.056	0.72	2.65
40	56	2.59	1.067	0.87	2.30
50	52	3.01	1.079	1.01	2.07
60	48	3.33	1.087	1.12	1.93
70	44	3.56	1.092	1.19	1.83
80	40	3.70	1.092	1.24	1.76
100	36	4.17	1.104	1.40	1.64

Fig. 1.26 Layout for flight model SOXE

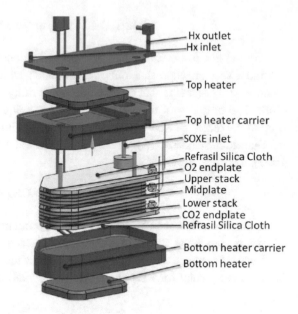

- Hx outlet
- Hx inlet
- Top heater
- Top heater carrier
- SOXE inlet
- Refrasil Silica Cloth
- O2 endplate
- Upper stack
- Midplate
- Lower stack
- CO2 endplate
- Refrasil Silica Cloth
- Bottom heater carrier
- Bottom heater

thermodynamic models, safe operating points (no carbon formation) were determined for each potential Mars gas flow rate. These are shown in Table 1.10.

1.4.2.6 Flight Model Assembly
The flight model include a heat exchanger to allow exiting hot gases to pre-heat incoming Mars gas, heaters at top and bottom of the stack, and electrical connections to the top, bottom and mid-point of the stack. See Fig. 1.26.

1.5 The Sabatier/Electrolysis Process

1.5.1 Introduction

In a system where hydrogen is brought from Earth, the Sabatier/Electrolysis (S/E) process occurs with hydrogen reacted with compressed CO$_2$ in a heated chemical reactor:

$$CO_2 + 4H_2 \Rightarrow CH_4 + 2H_2O \tag{1.1}$$

The reactor is simply a tube filled with catalyst. Since the reaction is quite exothermic, considerable waste heat is theoretically available for other purposes.

The methane/water mixture is separated in a condenser, and the methane is dried, and stored for use as a propellant. The water is collected, deionized, and electrolyzed in a standard electrolysis cell:

$$2H_2O + \text{electricity} \Rightarrow 2H_2 + O_2 \qquad (1.2A)$$

The oxygen is stored for use as a propellant and the hydrogen is recirculated to the chemical reactor. Note that only 1/2 as much hydrogen is produced as is needed for reaction [1.1], showing that an external source of hydrogen is necessary for this process to work. The only part of the S/E process that is unique and special is the catalyst particles used in the reactor. Twenty percent ruthenium on alumina (1–2 mm particle size furnished by Hamilton-Standard Co.) worked well as a catalyst (Zubrin et al. 1994; Clark 1996, 1997). The overall reaction is

$$CO_2 + 2H_2 \Rightarrow CH_4 + O_2 \qquad (1.2B)$$

An excess of methane is produced for the amount of oxygen produced. For each mole of oxygen produced, 2 mol of water must be electrolyzed.

If water is extracted from the Mars regolith as the source of hydrogen, Reaction (1.2A) is the starting point for Reaction (1.1). The overall reaction then becomes:

$$CO_2 + 4H_2O \Rightarrow CH_4 + 2O_2 \qquad (1.2C)$$

Note that the mixture ratio of CH_4 to O_2 is now very appropriate. Also note that 2 mol of water must be electrolyzed for each mole of oxygen produced.

The equilibrium mixture of molecules in a mixture of $CO_2 + 4H_2$ is shown in Fig. 1.27 as a function of temperature at a total pressure of 1 atm. As the temperature is raised, the equilibrium shifts away from the desired products of water + methane to $CO_2 + 4H_2$ but the rate of reaction increases. The challenge is then to operate the reactor at a temperature high enough that the kinetics are fast enough to allow a small compact reactor, yet the temperature is not so high that the equilibrium shifts too far to the left, leading to inadequate product yields. It has been found experimentally that at a reactor pressure of the order of ~ 1 bar, if a mixture of $CO_2 + 4H_2$ enters a packed bed of catalyst, a temperature near $\sim 300\ ^{\circ}C$ is high enough to approach equilibrium in a small reactor, and the equilibrium is far enough to the right that yields of over 90% $CH_4 + 2H_2O$ are obtained. If the exit zone of the reactor is allowed to cool below 300 °C, the yield can be >95% (Clark 1996, 1997).

There are two problems in the use of the S/E process that are closely coupled. The primary problem is the requirement to use hydrogen as a feedstock, necessitating either bringing hydrogen from Earth, or obtaining water from the subsurface of Mars. This is the main detraction of the S/E process. Transporting hydrogen to Mars and storing it for the duration of the process is difficult and problematic (see Appendix I). Obtaining water from the subsurface of Mars requires a significant campaign to prospect for water deposits and engineering to acquire the water (see Sect. 1.6). Closely related to this is the second problem with the S/E process; namely that it produces one molecule of methane for each molecule of oxygen, for an O_2/CH_4 mass ratio of 2:1 while the ideal mixture ratio for

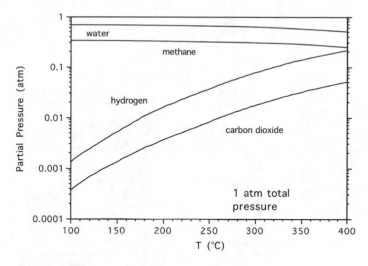

Fig. 1.27 Equilibrium mixture at 1 atmosphere in a Mixture of $CO_2 + 4H_2$

propulsion is roughly 3:1. There is a difference between the cases where hydrogen is brought from Earth versus hydrogen is obtained from indigenous water from Mars. When hydrogen is brought from Earth, an excess of methane is produced. But when hydrogen is obtained from Mars water, there is an additional (initial) step of water electrolysis that produces an extra molecule of O_2 per molecule of CH_4, so in this case, the O_2/CH_4 mass ratio is 4:1 and there is an excess of oxygen. Thus, there is an excess of methane for the amount of oxygen produced.

There are several possible approaches for transporting hydrogen to Mars and these are discussed in Appendix I. There are also various approaches to potentially reduce the hydrogen requirement. If a method (pyrolysis or partial oxidation) could be developed to recover hydrogen from the surplus methane, the hydrogen requirement might be reduced. Other approaches to reduce the hydrogen requirement include using an alternate conversion process that produces only oxygen (e.g. SOXE or RWGS) to augment the oxygen/methane ratio, or conversion of the methane to unsaturated hydrocarbons such as ethylene that have higher C/H ratios. These will be discussed in Sect. 1.5.3.

1.5.2   S/E Demonstration at LMA

With initial support from JSC (Zubrin et al. 1994; Zubrin et al. 1995) and subsequently from JPL (Clark et al. 1996, 1997), the S/E process for Mars was studied in some detail at Lockheed-Martin Astronautics (LMA) with a breadboard Sabatier-Electrolysis (S/E) demonstration unit that worked very effectively. The catalyst, electrolyzer and hydrogen recovery unit (from unreacted products) were supplied by Hamilton-Standard. A simplified diagram of the process is shown in Fig. 1.28.

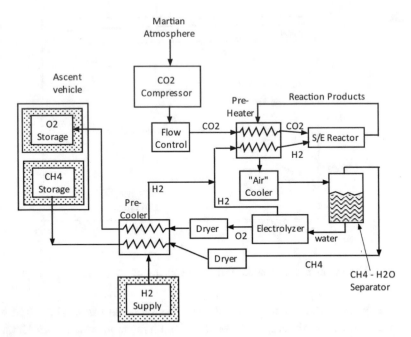

Fig. 1.28 Schematic flow diagram for Sabatier/electrolysis process. Cryocoolers are not shown, but are attached to each of the cryo tanks

It seems likely that this process could be made to work in a more compact, more autonomous design. The great advantage of this process is that it is well understood and seems to work well as a system with high conversion efficiency, good energetics and reliable start-up and shutdown capabilities. However, the entire system is quite complex in plumbing and components.

The description that follows is excerpted from the paper by Clark (1997):

Chemical conversion was completed in the Sabatier reactor where the CO_2 and H_2 mixture was exposed to a pre-heated catalyst bed. An advanced catalyst was used that offered nearly complete conversion with an efficient particle size (1–2 mm) using rough, irregular shapes for greater surface area. [It should be noted that initially, poor performance was obtained with larger and smaller catalyst particles, and it took quite a bit of trial and error to get the right catalyst for this application.] Heat from the reaction was routed to a prototype sorption-based CO_2 compressor by a combination of a coiled-tube heat exchanger that cooled the product stream and a copper braid attached to the reactor body to reduce the reactor outlet temperature. A backpressure regulator maintained reactor pressure and suppressed water vapor. A hydrogen recovery pump cleaned the hydrogen gas from the methane stream, enabling the recovery of nearly all of the excess hydrogen. The recovery pump could be used to enhance reactor efficiency by running the reactor rich with excess hydrogen to increase CO_2 efficiency, and then simply pumping the excess hydrogen from the product stream. The recovery pump could also be used to circulate hydrogen for warm-up and cool-down periods to minimize transient losses.

The Sabatier reactor was small, lightweight and required no power for continuous operation. An external electric heater provided startup power for pre-heat prior to operations. A highly active catalyst enabled self-sustaining operation while producing chemical conversion efficiencies of over 99% for both reactants when operating with the hydrogen recovery pump. Flow into the reactor was controlled with electronic flow controllers. These devices measured and controlled the flow based on set points generated by the computer control system. Two methods were used to recover heat from the reactor. The reactor used a copper heat conductor attached to the lower portion of the reactor body to conduct a portion of the heat away. The heat was applied to the base of the sorption compressor to generate carbon dioxide pressure. The heat conductor also lowered the reactor temperature at the outlet that increased the chemical efficiency. A coiled-tube heat exchanger recovered additional heat from reactor gas stream. Further cooling of the water/methane product stream was accomplished by a second heat exchanger prior to entering the water storage bottle. A backpressure regulator maintained the reactor pressure by regulating methane flow out of the water storage bottle.

Water flow into the electrolysis unit was controlled by a solenoid valve that opened when the electrolyzer water level dropped below a preset value. The electrolysis unit was a solid-polymer electrolyte membrane that generated high outlet pressures for hydrogen gas, eliminating the need for a mechanical compressor in the recycle loop.

Separation of the water into hydrogen and oxygen is a highly developed process. The electrolysis unit used for these experiments was supplied by Hamilton-Standard, and was a highly efficient and durable device based on space and military membrane and catalyst technology.

Water was separated from the hydrogen inside the integral tower attached to the unit. Oxygen was separated in a column on the outlet side of the electrolysis unit. A dryer removed residual water and carbon dioxide vapors from the oxygen stream. Zeolite was used in the dryer for increased efficiency and the ability to scrub the carbon dioxide.

A hydrogen membrane pump (supplied by the Hamilton-Standard Co.) recovered nearly all of the residual hydrogen gas in the methane stream for reuse by the reactor. The value of the pump was enhanced by the desire to operate the Sabatier reactor with an excess of hydrogen in order to almost entirely consume the carbon dioxide. This excess hydrogen could easily be removed from the product stream with the pump. Testing demonstrated several operating conditions that left the residual concentration less than 0.1% for hydrogen. The pump was capable of generating high pressures in the hydrogen output and feeds directly into the electrolysis hydrogen generator tower. A check valve isolated each pump and provided automatic pressure regulation that matched the hydrogen supply source.

In early testing, carbon dioxide was supplied to the S/E reactor from a tank of pure CO_2. Subsequently, all CO_2 was supplied from a sorption compressor that was basically a bed of Zeolite. The diurnal cycle was imitated by cooling down the compressor with cooling coils overnight, and high pressure CO_2 was generated during the day by heating the sorption bed. However, when this sorption compressor was exposed to a mixture of Mars gases (95% CO_2, with the remainder made up mainly of Ar and N_2) adsorption came to a stop fairly early, presumably due to buildup of a diffusion barrier by un-adsorbed Ar and N_2. This was alleviated by connecting a roughing pump to the sorption bed to keep gases circulating through the bed. This is an artificiality of the laboratory system and cannot be used in a real sorption compressor. Two possible approaches have been suggested for a flight system; one being use of a small fan or blower to continually circulate fresh Martian atmosphere through the sorbent bed and the other uses a small tank of high pressure carbon dioxide to periodically send a purge of high pressure gas through the sorption bed to flush out the permanent gases to the environment. Neither of these has yet been proven to be viable. Note added here: Use of a sorption compressor now appears to be superseded by cryogenic and mechanical approaches.

The system was operated for twelve days using the same flow rates and set points. Each day, methane and oxygen were produced for a minimum of six hours. All carbon dioxide was obtained from a sorption compressor (with the artifice of a roughing pump to remove un-adsorbed Ar + N_2) connected to a Mars atmospheric chamber containing a mixture of gases to simulate the composition of the Martian atmosphere at 6 Torr. A good consistency was found in the results with power consumption averaging less than 250 Watts and chemical conversion efficiencies well above 90%. Further full production days were completed while testing the hydrogen recovery pump. A demonstration run exceeded 99% efficiency for both reactants. With the residual CO_2 at less than one percent, the hydrogen pump had reduced the hydrogen concentration in the methane stream to less than one-tenth of one percent.

The CO_2 flow was held constant for all measurements and the hydrogen flow was varied to obtain various mixture ratios. A pressure of 1.5 atm in the reactor was used for all testing except a short series of pressure sensitivity test runs. A mass spectrometer continuously monitored the product stream. The CO_2 chemical conversion efficiency is the ratio of the methane percentage to the sum of the unreacted percentage plus the methane percentage. The hydrogen conversion efficiency is the ratio of methane percentage to the sum of {methane percentage + (hydrogen percentage)/4}. (In order to account for the molar ratio in the chemical equation, the hydrogen concentration is divided by four.) Figure 1.29 shows the hydrogen and CO_2 efficiencies from a series of test runs prior to extended duration testing. The lowest curve is the efficiency of the basic reactor with no hydrogen recovery. A large improvement is shown with the hydrogen recovery pump operating. Residual concentrations were reduced to less than 1% for each component at a mixture ratio of 4.7:1 while operating the hydrogen pump. The CO_2 conversion efficiency exceeded 99% when the mixture ratio was above 4.7:1 (stoichiometric = 4.0:1).

Testing indicated that reduced outlet temperatures increase the chemical efficiency. As the reactor outlet temperature increases, the conversion efficiency drops. The copper

Fig. 1.29 Measured Sabatier reactor conversion efficiencies for hydrogen and carbon dioxide (pump on/off refers to the hydrogen recovery pump)

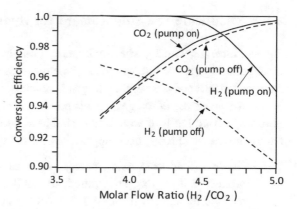

conducting braid to the sorption compressor reduces the reactor temperature by up to 50 °C, enabling continuous operations above 98% conversion efficiency.

The reactor processed ~ 30 g/h of CO_2 and produced ~ 21.5 g/h of O_2 and ~ 10.8 g/h of CH_4. The limiting factor in its capacity was the electrolyzer. The power required to run this system was mainly for the electrolyzer, which was ~ 125 W. A considerable amount of power was required to pre-heat the sorption compressor to supply CO_2 to the reactor, but once the process began, the exothermic reaction supplied most of the heat to the sorption compressor. However, after several hours of operation, as the sorption compressor reached higher temperatures, it required an increasing amount of power. (As stated previously, use of a sorption compressor now appears to be superseded by cryogenic and mechanical approaches.)

Holladay et al. (2007) demonstrated a microchannel apparatus for the Sabatier process. As we pointed out previously, it is not clear to this writer why anyone would want to use microchannel reactors for Mars ISRU processes, considering the high volumes of gases that must be throughput. Furthermore, despite the fact that almost any conceivable atmosphere acquisition system would have an elegant filtering system at the intake to reduce the likely possibility that some Mars dust might enter the reactor system, it would seem prudent to have sufficient reactor sizes so that minimal amounts of dust that do enter the reactor system can be accommodated. Nevertheless, Holladay et al. (2007) studied the Sabatier process in a microchannel reactor. They found that while equilibrium conversion should be high at 200–300 °C, "the kinetics of the reaction become very slow at these lower temperatures, requiring in increased contact time to reach close to equilibrium conversion. The net result is a requirement of larger reactor". By controlling the reactor to operate approximately isothermally, they were able to obtain 80% conversion at 400 °C. However, "over the course of several days of experimentation, there was a steady decline in the catalyst activity".

Comment: Incredibly, the paper by Holaday et al. did not refer to the pioneering work of Zubrin and Clark who obtained higher conversion efficiencies at lower temperatures ten years earlier!

1.5.3   Reducing the Requirement for Hydrogen in the S/E Process

As mentioned previously, when hydrogen is brought from Earth, the S/E process produces a methane-oxygen mixture with excess methane compared to an appropriate mixture ratio for a rocket. (But of course, if hydrogen is obtained from indigenous water on Mars, there is actually an excess of oxygen, so this problem disappears.) For the case where hydrogen is brought from Earth, it would be desirable to avoid the need for hydrogen to produce excess methane. Three general approaches for doing this are:

(i)   Use a second conversion process such as solid-state electrolysis or RWGS to produce half the oxygen required, so that the proper mixture ratio is obtained with half the throughput from the S/E process, thus halving the hydrogen requirement.
(ii)   Recover hydrogen from the excess methane produced by the S/E process.
(iii)   Convert the methane to a higher hydrocarbon or other organic with a lower H/C ratio than methane.

Recovery of hydrogen from methane can be approached by pyrolysis:

$$CH_4 \Rightarrow C + 2H_2 \tag{1.3}$$

by partial oxidation with oxygen:

$$2CH_4 + O_2 \Rightarrow 2CO + 4H_2 \tag{1.4}$$

or by partial oxidation with carbon dioxide:

$$CH_4 + CO_2 \Rightarrow 2CO + 2H_2 \tag{1.5}$$

Pyrolysis:

Pyrolysis is inherently a simple process:

$$CH_4(g) \Rightarrow C(s) + 2H_2(g) \; \Delta H = 74.9\,kJ/mol \tag{1.6}$$

Rapp et al. (1998) calculated the equilibrium concentrations of various species in the gas phase when methane is heated, as a function of temperature at a total pressure of 1 bar starting with pure methane. Their results for the mole fractions of the five major constituents are given in Table 1.11 and Fig. 1.30.

Clearly, the minor constituents can be neglected for our purposes. Since two molecules of hydrogen are produced for each molecule of methane that decomposes, the fractional conversion of methane to hydrogen is given by:

$$C = f\,H_2/(f\,H_2 + 2f\,CH_4)$$

Table 1.11 Mole fractions of constituents in methane pyrolysis

T (°C)	CH$_4$	H$_2$	C$_2$H$_2$	C$_2$H$_4$	C$_6$H$_6$	Conversion to H$_2$
400	7.75E−01	2.25E−01	4.69E−16	2.81E−09	6.58E−18	1.27E−01
500	5.09E−01	4.91E−01	1.89E−13	3.47E−08	3.06E−16	3.25E−01
600	2.49E−01	7.52E−01	1.56E−11	1.62E−07	3.19E−15	6.02E−01
700	9.97E−02	9.00E−01	4.35E−10	3.86E−07	1.18E−14	8.19E−01
800	3.94E−02	9.61E−01	5.82E−09	6.43E−07	2.49E−14	9.24E−01
900	1.68E−02	9.83E−01	4.72E−08	8.97E−07	3.89E−14	9.67E−01
1000	7.91E−03	9.92E−01	2.65E−07	1.13E−06	5.11E−14	9.84E−01
1100	4.06E−03	9.96E−01	1.12E−06	1.33E−06	5.92E−14	9.92E−01
1200	2.25E−03	9.98E−01	3.80E−06	1.48E−06	6.21E−14	9.96E−01
1300	1.32E−03	9.99E−01	1.07E−05	1.59E−06	5.75E−14	9.97E−01
1400	8.19E−04	9.99E−01	2.60E−05	1.65E−06	5.42E−14	9.98E−01
1500	5.29E−04	9.99E−01	5.55E−05	1.66E−06	4.61E−14	9.99E−01
1600	3.54E−04	1.00E+00	1.07E−04	1.62E−06	3.71E−14	9.99E−01
1700	2.45E−04	1.00E+00	1.87E−04	1.56E−06	2.85E−14	1.00E+00
1800	1.73E−04	1.00E+00	3.03E−04	1.47E−06	2.10E−14	1.00E+00

Fig. 1.30 Mole fractions in the conversion of methane to hydrogen

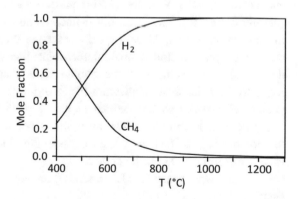

where the f's are mole fractions. It can be seen that at equilibrium, one can achieve 98% conversion of methane to hydrogen at 1000 °C and 99% at 1100 °C.

Our results differ significantly from those of Gueret et al. (1997). These authors found much lower conversions to hydrogen and much larger contributions of side reactions, with acetylene becoming a major constituent at higher temperatures. The disagreement is so strong that we suspect there is some error in their calculations. Our results indicate the equilibrium is very favorable to pyrolysis at 1000 °C and above.

The Hamilton Standard company developed a pyrolysis reactor that has the virtue that the carbon is formed as a single chunk of carbon with density roughly that of water rather than soot, so it does not create as much of a potential contamination problem (Noyes

1986, 1988). This system was developed for life support on Space Station. Hamilton Standard provided JPL with pyrolysis data taken in the 1980s. N. Rohatgi of JPL analyzed these data and estimated the dependence of the rate constant on temperature. He estimated that the rate is about 3 times faster at 1200 °C than it is at 1100 °C, and that a typical pyrolysis reactor is probably rate-limited rather than equilibrium limited, unless it has a long residence time. It is therefore likely that a pyrolysis reactor would be operated at >1150 °C to achieve a good reaction rate. Preliminary results taken at JPL in the 1990s confirm that temperatures of ~ 1150 °C or greater are desirable. A major challenge for pyrolysis is disposing of the carbon that is formed. Noyes and Cusick (1986) employed a quartz tube filled with quartz wool as a reactor, and they built up a "dense porous cake" of carbon on the quartz fibers as the reaction proceeded. (Work done in the 1990s at JPL shows that the carbon is deposited as a coating on the quartz fibers.) Their system was intended for manual removal of the carbon cake at periodic intervals on the space station. For a system on Mars, it does not seem practical to remove carbon cakes at intervals. For the near term, it is assumed that the accumulated carbon would have to be burned off by blowing CO_2 through a hot bed of carbon cake using the reaction:

$$C(s) + CO_2 \Rightarrow 2CO \ \Delta H = 172.6 \, kJ/mole \tag{1.7}$$

and venting the CO. Reaction (1.7) requires a good deal more energy than reaction (1.6) (the pyrolysis step). Therefore, if a method could be found to dispose of the carbon cake mechanically, that would be a great advantage. We are not aware that anyone has seriously attempted to find a method for doing this robotically. Until such a process is developed, we assume that for remote robotic applications, Reaction (1.7) will be required. The chemical equilibrium for Reaction (1.7) is shown in Fig. 1.31. High conversion efficiencies should be achievable at ~ 1000 °C if the rate proves reasonable. Work done at JPL in the 1990s (on minimal funding) provides a preliminary indication that the rate may be fast enough for consideration in a Mars ISRU system, but this needs further experimentation to be certain.

Note that when reactions (1.6) and (1.7) are combined, the overall reaction for the two steps is

$$CH_4 + CO_2 \Rightarrow 2CO + 2H_2 \ \Delta H = 247.5 \, kJ/mol \tag{1.8}$$

which is the same reaction used for methane reforming with CO_2. This shows that the energy required for the overall process is the same whether pyrolysis or reforming is employed, if the carbon cake is burned off to form CO in the pyrolysis process. If the carbon can be removed physically, the energy required for pyrolysis will only be 30% of that required for reforming.

It is possible that at intervals, one could pass CO_2 over the hot carbon to convert it to CO. It seems likely that a pair of reactors would be employed and one reactor would be

Fig. 1.31 Equilibrium mole fractions in the oxidation of carbon to CO by CO_2 (linear plot)

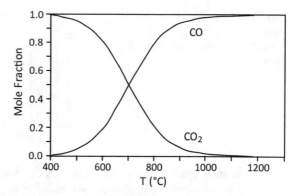

pyrolyzing methane during the day when power is available, while the other is burning off its carbon. These roles would be reversed every successive day.

In the 1990s, P. Sharma of JPL set up a laboratory facility to study both pyrolysis and reforming of methane with CO_2. The apparatus was designed to accommodate either reactor on a modular basis. The pyrolysis reactor was designed to accommodate a flow of 40 g/h of methane at a pressure slightly greater than 1 bar, and was designed to operate in the range 1000–1200 °C. For methane pyrolysis, the reactor basically consisted of a quartz tube packed with quartz wool. The quartz tube was placed inside an electrically heated furnace and maintained at a temperature in the range 1000–1200 °C. The gas exiting the reactor was expected to be mainly hydrogen (>98%) and was routed to the mass spectrometer for analysis. The same flow system was also used to test a methane-to-hydrogen reforming reactor.

Using rate constants derived from the Hamilton Standard data, N. Rohatgi of JPL roughly estimated the conversion and residence time as a function of position down the length of the reactor as shown in Fig. 1.32 for a steady state. The methane entering the reactor is gradually heated to about 1200 °C, after passing through a pre-heater that brings the entrance temperature up to 750 °C. Most of the reaction takes place in the second half of the reactor as temperatures approach 1200 °C. Based on these rough estimates it would appear that for a residence time of about 100 s, conversion of methane to >99% would be achieved in a 1.5 in. diameter, 12 in. long reactor. However, in reality, there will be no steady state, and the actual reaction zone will vary significantly with time as the carbon cake builds up. The actual rate of reaction will depend on heat transfer, reactor design and time. Nevertheless, these calculations provide an indication that good yields might be achieved in small pyrolysis reactors packed with quartz wool. The untimely death of P. Sharma prevented further development of this work, although NASA funding was drying up at that time anyway (as it usually does).

If pyrolysis is to become practical, it appears to require a second step in which reaction (1.5) is carried out. Clearly, the rate of this reaction will depend upon the porosity of the carbon cake, and will have to be determined experimentally. Therefore it would be necessary to carry out a simple test program to verify earlier results of Noyes and Cusick

Fig. 1.32 Calculated
conversion of methane along
the length of reactor assuming a
steady state

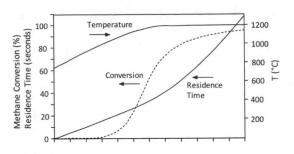

on yields in a quartz wool filled reactor in the 1000–1200 °C range. If high yields are
confirmed one could test to determine whether it is practical to carry out reaction (1.6) on
the carbon cake.

Recovery of Hydrogen From Methane by Partial Oxidation With Oxygen

Partial oxidation with oxygen appears to be a chemically viable process but it uses up
precious oxygen. Nevertheless, because it recovers hydrogen from two methane molecules
for each oxygen molecule that is expended, the possibility remains for using this hydrogen
to produce two oxygen molecules via the S/E process:

$$2CO_2 + 4H_2 \Rightarrow 2CH_4 + 2O_2 \qquad (1.9)$$

The question is whether enough is gained by recovering hydrogen from two methane
molecules for each oxygen molecule that is sacrificed.

Partial oxidation with oxygen would then work as follows. The S/E process would be
used to produce equal numbers of moles of oxygen and methane:

$$4CO_2 + 8H_2 \Rightarrow 4CH_4 + 4O_2 \qquad (1.10)$$

Now suppose we take a fraction z of the methane and a fraction $0.5z$ of the oxygen
produced by the S/E process and carry out a partial oxidation:

$$4zCH_4 + 2zO_2 \Rightarrow (4yz)CO + (8yz)H_2 \qquad (1.11)$$

with a conversion efficiency of y. Discard the CO, and the $(8yz)$ moles of hydrogen are
again reacted via the S/E process as follows:

$$(4yz)CO_2 + (8yz)H_2 \Rightarrow (4xyz)CH_4 + (4xyz)O_2 \qquad (1.12)$$

where the S/E conversion efficiency is x. The net effect of reactions (1.10)–(1.12) is the
overall reaction:

$$(4 + 4yz)CO_2 + (8)H_2 \Rightarrow (4[1 - z] + 4xyz)CH_4 + (4[1 - 0.5z] + 4xyz)O_2 \qquad (1.13)$$

Thus the methane/oxygen molar ratio changes from 1:1 without oxidation to

$$(4[1 - z] + 4xyz)/(4[1 - 0.5z] + 4xyz)$$

with partial oxidation. The number of moles of hydrogen required to produce one mole of oxygen is:

$$8/(4[1 - 0.5z] + 4xyz)$$

The power required to drive this process is mainly in the electrolysis step of the S/E process and is proportional to $(4 + 4yz)$. Using only the S/E process, the methane/oxygen molar ratio is 1.0 and it takes 2 mol of hydrogen to produce one mole of oxygen. The ideal mixture ratio is about 0.57 mol of methane for every mole of oxygen, and if this could be achieved, the required amount of hydrogen would be 1.14 mol per mole of oxygen.

The ability of partial oxidation with oxygen to reduce the hydrogen requirement is limited because oxygen must be used up in order to oxidize the excess methane. Figure 1.33 shows the dependence of the methane/oxygen molar ratio on the fraction of methane oxidized (z) for various assumed yields of the oxidation reaction (y), assuming $x \sim 0.95$. Although it is possible to reduce the methane/oxygen ratio below 1.0, it cannot reach the ideal figure of 0.57. Figure 1.34 shows the moles of hydrogen required per mole of oxygen produced, and again the ideal of 1.14 cannot be reached, although this ratio can be reduced well below the value 2.0 for the S/E process alone. It should be noted that the power required will be roughly proportional to $(4 + 4yz)$, and therefore the relative power required will be higher as one moves to the right or downward in Figs. 1.33 and 1.34. For a representative case, if $y = 0.9$ and $z = 0.6$, the power required will be 54% higher than for the S/E process alone. This would reduce the hydrogen requirement by about 17% from 2.0 to 1.65 mol of hydrogen per mole of oxygen. Even if the values of x and y were both 1.0, it would still take 1.54 mol of hydrogen to produce 1 mol of oxygen if $z = 0.6$, and thus it would require a 54% increase in power to reduce the hydrogen requirement by 23%.

It would appear that partial oxidation of methane using some of the oxygen produced by the S/E process is probably not an attractive proposition.

Recovery of Hydrogen From Methane by Reforming With CO_2

Reforming of methane with carbon dioxide would be a feasible process if the yields can be made high enough, although the power requirement is significant. There is a ready supply of carbon dioxide available from the Mars atmosphere compressor and using regenerative heat recuperation where the exit gases from the reactor heat the incoming gases should help. It should be possible to recover most of the hydrogen produced by carbon dioxide oxidation. Separation of the hydrogen from $CO + H_2$ mixture resulting from the reaction:

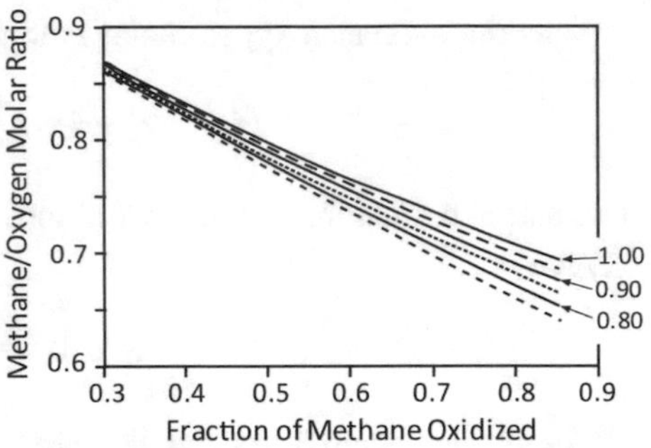

Fig. 1.33 Molar methane/oxygen ratio assuming x = 0.95

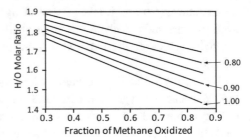

Fig. 1.34 Molar hydrogen/oxygen ratio assuming x = 0.95

$$CO_2 + CH_4 \Rightarrow 2CO + 2H_2 \tag{1.14}$$

is the main challenge.

An analysis of reaction (1.13) was made by Rapp et al. (1998). The theoretical equilibrium conversion depends primarily on 3 factors: temperature, reactor pressure, and the CO_2/CH_4 molar reactant ratio. The conversion increases with increasing temperature, decreasing total pressure, and increasing CO_2/CH_4 molar reactant ratio. In addition, if hydrogen is removed from the reactor by a hydrogen permeable membrane, the partial pressure of hydrogen in the reactor will be lowered, driving the reaction to the right at lower temperatures than would be required if equilibrium prevailed.

Figure 1.35 shows the fraction of methane converted to hydrogen for a CO_2/CH_4 molar reactant ratio = 1.5 as a function of temperature at several reactor pressures. It can be seen that >90% conversion can be achieved at 900 K for a CO_2/CH_4 molar reactant ratio = 1.5. As the CO_2/CH_4 molar reactant ratio is increased (decreased) the conversion curves shift to lower (higher) temperatures. The effect of removal of hydrogen from the reactor is shown in Fig. 1.36 for various percentages of removal of hydrogen from the reactor starting with 1.5 mol of CO_2 for each mole of CH_4 at a total pressure of 0.5 bar as a function of temperature. If 90% of the hydrogen formed can be removed from the reactor at 850 K, this allows the same conversion of methane as the equilibrium value at 950 K.

Fig. 1.35 Methane conversion factor in the $CO_2 + CH_4 \Rightarrow 2CO + 2H_2$ reaction starting with 1.5 mol of CO_2 for each mole of CH_4

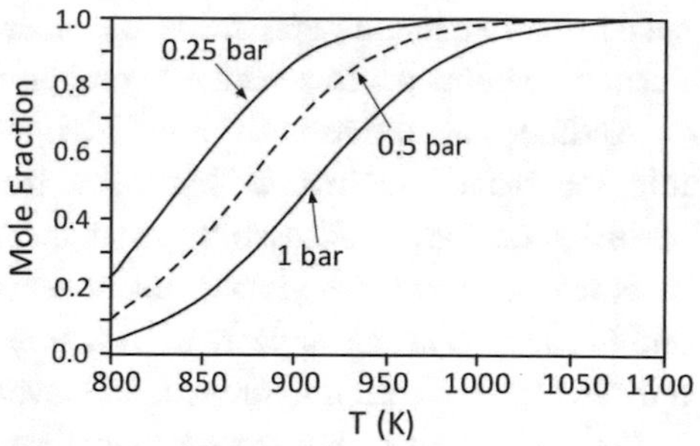

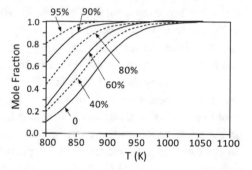

Fig. 1.36 Methane conversion in the $CO_2 + CH_4 \Rightarrow 2CO + 2H_2$ reaction starting with 1.5 mol of CO_2 for each mole of CH_4 at a total pressure of 0.5 bar for various percentages of removal of hydrogen from the reactor

This process was studied by a quite a number of investigators. Rapp et al. (1998) provided a detailed review and analysis of the literature. A variety of catalysts have been tested and some progress was made on elucidation of the mechanism of reaction. The most serious problem encountered is gradual buildup of carbon on the catalyst, leading to reduced performance or even deactivation. Platinum on YSZ had very low rates of carbon formation. In general, a catalyst coated with carbon can be regenerated by running the reactor for a while with pure CO_2 to oxidize the carbon.

Experiments by van Keulen et al. (1997) showed that stable $\sim$90% conversion can be achieved in a reactor at 800 °C using a Pt/ZrO_2 catalyst. Mark and Maier (1996) found excellent results with Rh catalysts supported on either γ-alumina or ZrO_2 (5%)/SiO_2. These catalysts showed high activity and stable performance after long reaction times. They also showed a significant benefit to adding extra CO_2 in the feed. The conversion of CH_4 increased significantly as the CO_2/CH_4 ratio was increased. It appears likely that high yields can be obtained under some conditions for the reforming of methane into hydrogen using carbon dioxide. The question then arises as to how to separate the hydrogen and compress it for recirculation to the Sabatier reactor. One could attempt to cool down the reaction products, which would be roughly 50% CO and 50% H_2, and remove the hydrogen from this mixture. Nafion membranes can separate hydrogen from gas mixtures and pressurize the hydrogen in one step (Sedlak et al. 1981) but Nafion membrane systems tend to be poisoned by even small amounts of CO. However, the Hamilton Standard Company indicated that it may be able to develop a CO-tolerant Nafion membrane system. Other membranes made by other manufacturers might also work. Zubrin used Permea membranes to separate hydrogen from hydrogen-CO mixtures.

Another alternative is to insert high-temperature membranes (such as Pd coated Ta) into the reactor to transfer hydrogen from the reactor to an outside reservoir. Some investigators have successfully used membrane reactors to remove hydrogen from the reactor to enhance the yield at lower temperatures. (Ponelis and van Zyl 1997; Kikuchi and Chen 1997). This appears to be a potential way to both increase the yield and provide a means for separating hydrogen. However, since one requires a lower hydrogen pressure in the tubes than in the reactor, and furthermore removal of hydrogen from the reactor lowers the partial pressure of hydrogen in the reactor, the pressure of hydrogen in the tubes leading away from the Pd/Ta tubes will be relatively low. Some form of compressor would have to be connected to the outlet of the Pd/Ta tubes to establish a low hydrogen pressure in the tubes while delivering hydrogen at >1 bar back to the remainder of the ISPP system. One approach would be to use a Nafion solid polymer membrane, such as that developed by the Hamilton-Standard Co. This pump requires that the total pressure on the cathode side be around 0.3 bar or greater, and since the hydrogen partial pressure would probably be around ~ 7 Torr, it is necessary that an inert "carrier gas" be included in the tubes. Since the Nafion pump transmits a small amount of this carrier gas in addition to the hydrogen, the carrier gas will gradually become depleted and must be replaced at infrequent intervals. Unfortunately, hydrogen diffusion through the carrier gas is slow and one would have to rapidly recirculate carrier gas in a loop to transport hydrogen from the reactor tubes to the Nafion membrane. The power requirement for such a system will be significant.

Yet another approach is to use an adsorption system to create a low hydrogen pressure in the tubes carrying hydrogen from the reactor. Such a system would require two beds of sorbent materials. One bed would be maintained at Mars ambient temperature and would adsorb hydrogen as it appears in the tube connected to the reactor. If the proper adsorbent is used, it will establish a hydrogen pressure in the 5–10 Torr range at Martian temperatures. At the same time the other bed is isolated and heated electrically (or possibly with use of hot gases exiting the reactor) to evolve hydrogen at high pressure (1–2 bar) for recirculation to the Sabatier process. These two adsorption beds are operated in tandem, and their roles are reversed every successive day by opening and closing valves. A number of materials have been developed over the past decade or two for adsorption compression of hydrogen over various temperature and pressure ranges. Mg_2Ni appears to fit this need best. It can absorb at low pressures around room temperature and provide 2 bars of hydrogen pressure at around 530 K, and it can deliver about 3 g of hydrogen per 100 g of Mg_2Ni. Since a full scale system is likely to need to process about 70 g of hydrogen per 7 h operational day, each sorbent bed must have the capacity to store ~ 70 g of H_2. If the storage capacity is 3%, this would imply that each bed would contain ~ 2.1 kg of sorbent. Such sorbent beds have been operated routinely for years. The main difficulty in their use is the possibility of contamination by impurities. The practicality of such a system requires further evaluation.

Finally, a small mechanical compressor might suffice.

Conversion of Methane to Higher Organic Molecules

A wide variety of processes for conversion of methane to higher organic molecules have been proposed. Obviously, conversion to ethylene, or even ethylene/ethane mixtures, would have great value in both reducing the hydrogen requirement (ethylene requiring two hydrogen atoms per carbon atom compared to methane which requires four) as well as in eliminating the need for cryogenic storage of methane. Other possibilities include conversion to benzene, dimethyl-ether, or alcohols.

In general, most processes for conversion of methane to higher organics suffer from low selectivity of products at high yields and low yields at high selectivity. There doesn't seem to be any straightforward process for converting methane to higher organics that produces high yields with high selectivity.

Most processes utilize oxygen in some form as an intermediary and a fundamental problem is due to the fact that ethylene and ethane are far more reactive with oxygen than methane and therefore are easily oxidized to CO/CO_2 when their concentrations become comparable to that of methane. There seems to be a 30% conversion barrier in converting methane to ethylene. Makri et al. (1996) have shown that this 30% barrier can be broken with the effective use of molecular sieves.

In the oxidative conversion, the conversion of methane to ethylene appears to follow the following mechanism:

$$2CH_4 \underset{1}{\Rightarrow} 2CH_3 \cdot \underset{2}{\Rightarrow} C_2H_6 \underset{3}{\Rightarrow} C_2H_4 \underset{4}{\Rightarrow} 2CO_2$$

The above four-step mechanism shows that if the reaction is not arrested at step 3, and step 4 takes place, one would get a very poor conversion to C_2H_4. The catalytic system design should then be such that steps 1–3 are carried out efficiently, but step 4 is frozen. This can be accomplished by a reaction scheme that employs a molecular sieve based trap to remove (and thus protect) the ethylene before it can be oxidized. Such a scheme would avoid step 4 and thus provide a high C_2H_4 selectivity. Ethylene can be recovered later from the adsorbent trap by heating. This process also requires recirculation of reaction products so that more and more ethylene is trapped out on the sieve with each pass through the reactor.

Some catalysts can convert methane to ethylene and other higher hydrocarbons directly, without going through the oxidative route. The penalty is an increased energy requirement, due to an endothermic heat requirement of 221.5 kJ/mol (53 kcal/mol). Reasonably good selectivities for higher hydrocarbons have been obtained, but methane conversions were unacceptably low (around 10%).

Thermodynamic temperature limitations for methane conversion to higher hydrocarbons can sometimes be circumvented by using a two-step process. In this approach, a catalyst is chosen that can activate the CH_4 molecule (for C–C coupling) at a low temperature. If this step is rate-limiting, the actual coupling to higher hydrocarbons occurs in the second step. In practice, this approach has had only a partial success.

Conversion of methane to benzene has been studied in detail by several investigators. In general, selectivities of 65–90% have been obtained, but conversion yields were only in the 7–12% range. The prospects for conversion of methane to benzene do not look too good because of low yields.

After reforming methane with carbon dioxide to produce carbon monoxide and hydrogen, further reactions can lead to higher organic products. For example the Fischer-Tropsch synthesis involves conversion of CO and H_2 to straight-chain alkanes:

$$(2n\ +\ 1)\,H_2 + n\ CO \Rightarrow C_nH_{(2n+2)} + n\,H_2O$$

However, a number of side reactions can take place, and one may obtain alkenes and alcohols in addition to the alkanes. High conversion with high specificity does not seem to have been achieved.

Another possibility is to catalytically react CO with hydrogen to produce dimethyl ether. This low-boiling liquid has a H/C ratio of 3, compared to methane that has a H/C ratio of 4, and utilizes more of the oxygen bound in the starting material, carbon dioxide.

The overall reaction is:

$$2\,CO + 4\,H_2 \Rightarrow H_3COCH_3 + H_2O \quad \Delta H_{298} = 205.5\,kJ/mol$$

1.6   Obtaining H$_2$O on Mars

If indigenous water can be obtained on Mars, that would provide many benefits. It would allow production of both CH_4 and O_2 propellants, not only for ascent, but also to power pressurized rovers, and also to provide a portable power source for other site operations. It would also provide a significant augmentation and backup for the recycling system.

The question arises as to the practicality of obtaining indigenous water on Mars. Abbud-Madrid et al. (2016) studied this question in the so-called "M-WIP Study".

The objectives of the study were to:

(1)   Formulate descriptions of hypothetical H_2O reserves on Mars,
(2)   Estimate the rough order-of-magnitude of the engineered system needed to produce water for each of four reference cases,
(3)   Prepare a first draft analysis of the sensitivity of the production system to the known or potential geological variations, and
(4)   Derive preliminary implications for exploration.

Four reference cases were defined:

Case A—glacial ice.

Case B—a natural concentration of poly-hydrated sulfate minerals.

Case C—a natural concentration of phyllosilicate minerals.

Case D—regolith with average composition as observed from in situ missions.

They estimated the requirements for producing 28 mT of O_2 and 7 mT of CH_4 for Cases B, C and D. These are shown in Table 1.12, assuming 100% efficiency in water extraction. Actual regolith processing requirements will undoubtedly be greater.

The study did not clarify the requirements for mining ice from regolith at higher latitudes.

In their conclusions they said:

Whether any of these cases is above minimum thresholds for a potential future human mission depends on the resource envelope for that mission, as well as its architecture and priorities—none of which has yet been determined.

The different cases have different sensitivity to known or potential natural geologic variation. The granular materials cases (B-C-D) are most sensitive to the nature/scale of the mechanical heterogeneity of the ore deposit, and the distance between the mine and the processing plant. The ice case (A) is most sensitive to the thickness and properties of the overburden.

We do not have enough orbital or ground data to be able to determine if deposits as good or better than the reference cases exist at Mars. Exploration is needed at several different scales. The details of the logic imply that this is a 2-step exploration problem—there needs to be an orbital reconnaissance mission followed by at least one landed exploration mission.

Follow-up work is needed in multiple areas, including technology development for ice and granular mining cases, advance mission planning …

Cases C and D are clearly very inefficient compared to Case B. It seems evident to this writer that within the realm of tapping granular materials, Case B is the only way to go—provided sites can be found with good gypsum concentrations at low elevations and acceptable latitudes. There is some evidence of gypsum-rich areas on Mars, as for example given by Szynkiewicz et al. (2010), and others.[5] Gypsum-rich sites at high northern latitudes would not be of interest since ready ice would be available. This entails prospecting for sites rich in gypsum, and balancing out site selection with factors other than ISRU for the Mars mission. The only virtue in Cases C and D is that they occur almost everywhere and do not restrict site selection.

Hoffman et al. (2016) wrote an "addendum to M-WIP study, addressing one of the areas not fully covered in this report: accessing and mining water ice if it is present in certain glacier-like forms". While the M-WIP study assumed production of 16 mT of water to produce 28 mT of O_2 and 7 mT of CH_4 for ascent propellants, Hoffman et al. proposed 20 mT of water per crew landing ("each crew's ascent vehicle plus a nominal amount that is typically consumed in EVA cooling and other miscellaneous ECLSS losses") and over all, 100 MT of water for five successive crew landings. They proposed

---

[5]https://themis.asu.edu/feature/32, http://www.elementsmagazine.org/archivearticles/e6_2/king.pdf, https://websites.pmc.ucsc.edu/~fnimmo/website/abstracts/bishop.pdf.

storing 20 mT of water in a bladder 8 m indiameter and 0.4 m thick. They did not discuss thermal management of the water that likely would be frozen.

Hoffman et al. (2016) divided the Mars surface into five zones:

(1) Variable 45° to 50°N up to 90°N latitude: semi-continuous shallow ice within 0.3 m of surface.
(2) Variable 25°N to 30°N latitude: discontinuous ice within 5 m of surface.
(3) Variable 30°S to 25–30°N: No shallow ice suspected within 5 m of surface.
(4) Variable 30°S to 45–50°S latitude: discontinuous ice within 5 m of surface.
(5) Variable 45–50°S to 90°S latitude: semi-continuous shallow ice within 0.3 m of surface.

As they pointed out,

> Landing and ascent technology options place boundaries on surface locations leading to a preference for mid-to low-latitudes and mid-to low-elevations. Accessing water ice for … ISRU purposes is attractive, leading to a preference for higher latitudes if water ice is the desired feedstock.

In their study, they placed preliminary latitude boundaries at ±50°.

Hoffman et al. presented a model that showed that the required propellant mass for ascent increased with increasing launch latitude. In going from the equator to 50° latitude, the total lift off mass increased by about 4.5%. However, if all ascent propellants are produced by ISRU, this might not matter much.

A preliminary elevation boundary was set at no higher than +2 km (MOLA reference). Most of zone (1) lies too high to fit this criterion. Most of zone (2) lies lower than 2 km.

Hoffman et al. asked an interesting question: If the putative ice is covered by a thick overburden of regolith, at what depth of ice will removal of the overburden total up to moving the same amount of regolith as would be required for processing regolith in cases (B), (C) and (D)? They estimated the following:

Over burden thickness (m)	Equivalent to
2.2	Case B
3.7	Case C
5.5	Case D

We may conclude that if the overburden is greater than about 2 m, the amount of regolith to be exhumed in extracting ice might be comparable in extent to that for processing gypsum-rich regolith. Hoffman et al. discussed approaches for extracting ice from overburden.

Hoffman et al. pointed out that one approach involves drilling through the overburden and melting ice in situ, pumping liquid water upward. The approach of melting the ice in situ has been used in terrestrial applications a number of times. A cased hole is required

Table 1.12 Requirements for obtaining 16 mT of water from various Mars soil sources

Source	Water content (%)	Extraction temperature (K)	Mass of regolith processed (mT)
B: Gypsum enriched regolith	8.6	575	390
C: Smectite clay enriched regolith	2.7	575	1200
D: Typical Mars regolith	1.3	425	2600

so that the cavity can be sealed and pressurized to minimize water sublimation. As they pointed out, such systems were used over the past fifty years at Greenland and Antarctica.

Rucker (2013) carried out a study of drilling techniques for Mars but it was not specifically addressed to obtaining water for ISRU. A drill is needed to penetrate the overburden and a casing must be emplaced to seal the hole. A system is needed to melt ice in situ and pump meltwater upwards.

The conclusion we must draw at this early stage is this. It is unlikely that the landing site will be at a higher latitude than 50°N. That being the case, the probability of finding near-surface ice deposits is low. While it is possible that near-surface ice might exist in some locations at latitudes near or slightly below 50°N, locating them and validating them will take considerable effort. At higher latitudes, near-surface ice should be readily available but as we pointed out, it is very unlikely that a landing would take place there for other reasons. Hence the case for obtaining water directly from near-surface ice in a human mission to Mars is not strong at this point.

Kleinhenz and Paz (2017) reported on a study that built upon previous M-WIP work. They claimed:

> Since the publication of DRA5.0, robotic and orbital exploration of Mars have indicated a greater likelihood, and more prevalent presence, of water in the Mars regolith.

Based on that assumption, they proceeded to outline an ISRU system that utilized indigenous Mars water. However, it is not clear to this writer what new exploration results (post-DRA 5.0) were the basis for the claim that the case for Mars water is now stronger.

In one part of the report, Kleinhenz and Paz (2017) discussed an excavation system that seemed to be addressed to water of hydration in regolith. This was based on the so-called RASSOR rover. This is a small rover (mass ∼66 kg) that uses twin bucket drums to excavate ∼80 kg of regolith. In one approach, the excavator would deliver regolith to, and from a centralized ISRU processing plant. Alternatively, a mobile processing system could be used in which the regolith is processed at the excavation site and water is delivered to the ISRU system. However, a more capable mobility platform than the RASSOR would be required for the mobile processing system option. The rover is assumed to travel ∼100 m to retrieve fresh material. According to the article:

A single rover trip involves excavating fresh regolith and delivering it to the ISRU system, as well excavating the spent regolith and depositing it 10 m away from the excavation site. RASSOR is battery powered so the model accounts for the timeline balance between resource range, battery recharge time, and number of trips needed to meet production requirements. More excavators are added as needed to meet these requirements. Laboratory tests in regolith simulant provided parameters such excavation time, speed, and power consumption during both traverse and excavation.

To retrieve about 20 mT of water from regolith containing 1.3% water, assuming 100% efficiency, will require processing 20/0.013 = 1540 mT of regolith. With each load amounting to 0.08 mT of regolith, a total of 1540/0.08 = 19,250 loads would need to be carried. Although the report mentioned "time line" several times, no specific information was presented. A wild guess is that the process of excavation, delivery of regolith to processor, disposal of spent regolith, and recharge of batteries takes at least one hour, probably several hours. That would imply that the RASSOR is undersized for a 14-month ISRU sequence.

The processor was described as follows:

The regolith processor subsystem model consists of three key components, a regolith dryer, vapor cleanup and cold trap. The regolith dryer is based on published experimental data of screw conveyor dryers designed to remove moisture from large quantities of granular materials. The vapor cleanup component is based on COTS hardware that has proven to be effective at separating contaminants from water vapor. The coldtrap model is designed to calculate the cold surface area required to condense a given amount of water by providing an assumed heat transfer coefficient and temperature at the cold surface. The surface area takes the form of semi-circular fins, and a spherical container surrounding the fins is used as the pressure vessel and water reservoir.

(As usual in NASA studies, they assumed the ISRU system would operate for 16 months, whereas we have pointed out that it is more likely that ISRU operation would be restricted to about 14 months.)

In continuing their analysis Kleinhenz and Paz focused on Case D (rather than B or C) because it is applicable over the broadest range of surface. They estimated requirements for thermal power and mass of regolith processed as shown in Fig. 1.37.

We can check these figures, starting with mass of regolith processed. We assume that 20 mT of water must be produced. For Case (D) with 1.3% water content, the amount of regolith that must be processed (assuming 100% extraction) is 20/0.013 = 1540 mT. For Case (C) with 2.7% water, the amount of regolith that must be processed (assuming 100% extraction) is 20/0.027 = 740 mT. And for Case (B) the amount of regolith that must be processed (assuming 100% extraction) is 20/0.086 = 230 mT. Evidently, the data in Fig. 1.37 are too low.

Next consider thermal power. The amounts of regolith previously given must be processed over a time period of 14 months = 10,080 h = 6.05×10^5 s. Assuming the regolith specific heat is 1.5 J/°C-g, and a Mars temperature of −80 °C, we obtain the data

Fig. 1.37 Estimated power and regolith mass (Kleinhenz and Paz)

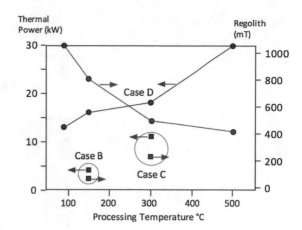

Table 1.13 Estimated power requirements for processing regolith

Case	Regolith mass (g)	Total energy over 14 months (J)	Processing temperature (°C)	Power (J/s) = W	Power (kW)
B	2.3×10^8	7.9×10^{10}	150	13,200	13
C	7.4×10^8	4.2×10^{11}	300	70,000	70
D	1.7×10^9	9.7×10^{11}	300	163,000	163
D	1.5×10^9	1.3×10^{12}	500	217,000	217

shown in Table 1.13. Evidently the data provided by Kleinhez and Paz are massive underestimates. While Kleinhez and Paz said

> The focus was primarily case D, which is a mixture of many different phases and therefore has a broader release profile.

But as Table 1.13 shows, Case D has impractical power requirements. The results of Kleinhez and Paz appear to be anomalous. Calculations presented here confirm that Case (D) is thoroughly impractical and Case (B) is eminently practical.

1.7 Obtaining Water from the Atmosphere

Toward the end of the 20th century, Professor Adam Bruckner and his students at the University of Washington developed a concept for adsorbing water vapor out of the Martian "air" (e.g. Williams et al. 1995; Coons et al. 1995, 1997). This water vapor adsorption reactor (WAVAR) is conceptually very simple, as can be seen from Fig. 1.38. Martian atmosphere is drawn into the system through a dust filter by a fan. The filtered gases are passed through the adsorbent bed, where the water vapor is removed from the flow. Once the bed reaches saturation, it is regenerated and the desorbed water vapor is

Fig. 1.38 WAVAR system

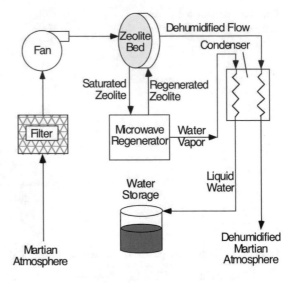

condensed and piped to storage. The design has only six components: a filter, a fan, an adsorption bed, a regeneration unit, a condenser and an active-control system. For robotic applications, solar power would only be available for limited periods during the day, so the WAVAR would be designed for absorbing wind-driven atmosphere at night, and using the blower and regenerator only when power is available during the day.

The WAVAR fan has to move a large volume of low-temperature (200–240 K), low-pressure (~6 Torr) gas. Adsorption is used to concentrate water vapor on a fixed bed. Because the adsorbence of a bed is finite, the adsorption-desorption cycle is necessarily a batch process. For water vapor adsorption on Mars, the choice of adsorbent is limited to those with an aperture of 3 Å (slightly larger than a water molecule), such as Zeolite 3 A. In the original WAVAR concept, the pelletized adsorbent was packed into a disc-shaped bed (Fig. 1.39) that was divided radially into sectors. Each of these sectors was sealed from the others with insulated dividers to prevent lateral heat and mass flow during regeneration. The bed was rotated stepwise, with one sector at a time brought into the regeneration unit while the others were adsorbing water from the flow. The bed's rotation was timed by the control system so that the sector about to enter the regeneration unit was just reaching saturation. The control system monitors the humidity of the output and adjusts the rotational rate and regeneration time for maximum efficiency.

Regeneration, or desorption, of the bed is achieved by thermal swing regeneration that involves heating the bed to drive off the water. The thermal swing can be accomplished either through resistive heating or with microwaves. The major advantage of using microwave energy over conventional conductive heating is that it provides rapid uniform heating for reduced regeneration time and can be tailored to specifically heat water molecules. In the original WAVAR concept, the microwave regeneration unit remained in place while the adsorbent bed rotated through it stepwise (Fig. 1.40). A saturated section

Fig. 1.39 WAVAR Zeolite
bed and regenerator (original
design)

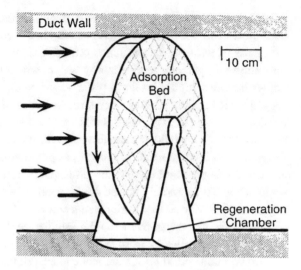

Fig. 1.40 WAVAR adsorbent
wheel for original concept

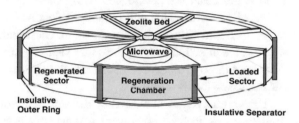

is rotated into the regeneration unit. Each sector is isolated with insulated separators and an insulated end-cap. The sealing mechanism of the regeneration unit consists of two plates with Teflon seals that engages the separations and end cap, forming a sealed chamber for regeneration.

In later work, a new geometry was designed with the sorbent arranged peripherally around the circumference of the wheel.

The WAVAR approach is highly dependent on how much water vapor is available in the Martian "air." One rough estimate that has been used by some is that the global average concentration of water vapor is about 0.03%, as given by Carr in his book on Mars water. Yet this data is based on ephemeral support. Furthermore, the water vapor concentration varies significantly with location, and at very high northern latitudes in local summer, can be many times higher than the global average. Also unknown is how the water vapor concentration varies with height above the surface of Mars. Adan-Plaza et al. (1998) used Viking data for local humidity at the two sites but the range of values was so large as to make it essentially impossible to pin down a water concentration. A number of studies have indicated that the total column abundance of preciptatable water in middle latitudes, while variable, can be roughly estimated to be about 15 μm. If we consider a vertical column of cross sectional area 1 cm^2 and height 10 km, with surface pressure

5 Torr, but the pressure decreases with height, and the average pressure over the whole column is about 3.6 Torr. The mass of CO_2 in the column is then estimated to be 11.6 g. The mass of water in the column is 0.0015 g. If the water were uniformly distributed in the column the percentage of water by weight would be $100 \times 0.0015/11.6 = 0.013\%$. At higher latitudes, higher levels of water vapor would be expected.

WAVAR requires moving very large volumes of "air" though the adsorbent per day per kg of water per day. For robotic applications where the Lander depends on solar power, WAVAR would presumably operate entirely in the adsorption mode at night, depending on Martian winds to blow atmosphere through the sorbent bed. During the day, an active blower would drive the atmosphere through the bed, and each pie-shaped segment would be serially heated to drive off water. However, the possibly high power requirement is viewed as a serious concern.

One of the first things that should be studied in the WAVAR concept is whether water vapor can be selectively adsorbed from Mars gas containing perhaps 0.013% water vapor at ~ 5 Torr and ~ 200 K. If a sorbent can adsorb most of the water vapor in a reasonable contact time, and not get flooded with excess CO_2, then WAVAR would appear to be technically feasible, although other considerations might make it impractical. It is not a trivial matter to make such a measurement. One would have to flow CO_2 at 5 Torr through a bed of ice at 200 K and acquire water vapor by sublimation. This, in turn would have to be passed through a bed of sorbent, which can eventually be isolated and heated to drive off the water that is adsorbed.

The various papers published by the University of Washington group at the end of the 1990s provide analyses of slow rates, power needs and other vital aspects of the WAVAR concept. It is unfortunate that funding for further research on adsorption of very small amounts of water vapor has not been available since then.

1.8   Ancillary Needs for Mars ISRU

1.8.1   The ISRU–MAV Connection

ISRU produces dry gaseous O_2 and the gaseous O_2 must be liquefied and eventually stored in Mars Ascent Vehicle (MAV). The physical layout of the ascent vehicle, the descent vehicle, the ISRU system, and the power system will be an important part of any Mars mission design, and these have not yet been explored thoroughly. One current view is that the descent vehicle would provide a platform on which the ascent vehicle would be mounted, with the ISRU system on a platform where the descent and ascent vehicles intersect. This might prove to be effective, but we are still at an early stage of mission design. One way or another, the ISRU system must supply liquid cryogenic propellants to the MAV.

Sanders (2016) reported on a NASA study of the ISRU–to–MAV interface and liquefaction processes. On the one hand it is desirable that the MAV be as light as possible.

This would augur to minimize insulation and use active cooling/liquefaction hardware. However, this will increase the heat leak and power needed to keep propellants liquefied, and it will increase the required surface power and radiator mass.

Three options were considered:

1. Batch liquid transfer from ISRU (located on Mars Descent Module, MDM) to MAV (Pros: All liquefaction hardware and power is located on MDM; Cons: Separate tank and pump and dedicated cryo-cooler, Transfer line must be re-cooled for each transfer, Ullage return line required.)
2. Steady liquid transfer from ISRU to MAV (Pros: Liquefaction hardware and power is located on MDM, Transfer line stays cold, No accumulator required on MDM; Cons: Dedicated cryo-cooler required, Ullage return line required).
3. Steady O_2 gas transfer to MAV and liquefy in MAV Tanks (Pros: Simplest system, No cryo-cooler/liquefaction hardware on the MDM, No accumulator on MDM, Only one transfer line and quick disconnect for MAV launch; Cons: Significant additional cooling capability required in MAV tanks, Approximately 400 W cooling for 14 month fill cycle).

These options are now being studied further by NASA.

The ISRU system and the liquefying system generate a good deal of waste heat. Large radiators will be required. Studies are now underway to quantify this. Total heat rejection from ISRU, liquefaction and MAV "keep alive" was estimated to be 25 kW. Initial estimates suggested a radiator area of 460 m², but later refinements indicated that as little as 200 m² might suffice.

Sanders (2016) also briefly described other applications on the Mars surface in an "ISRU-rich environment" including life support backup and mobile power.

1.8.2 Power System

1.8.2.1 Power Requirements at Full Scale

The power requirement for CO_2 acquisition and compression appears to be the lowest for the mechanical pump option. As we showed in Sect. 1.3.1.3, the MOXIE compressor entails a power requirement of roughly 80 W at a plenum pressure of 150 Torr. The MOXIE flow rate is roughly estimated at ~50 g/h.

Table 1.5 shows the oxygen production rates needed for crews of 4 and 6. The corresponding CO_2 flow rates for solid oxide electrolysis production of oxygen were shown to be as follows.

The required CO_2 delivery rate for a crew of four is about 2.3 kg/h $\times$ (88/32) $\times$ (1/0.5) = 12.7 kg/h. For a crew of six it is about 19 kg/h. If some of the pressurized CO_2 in the electrolysis tail gas can be recovered and recirculated, this would be reduced, possibly significantly. For our purposes here we will assume the required CO_2 flow rate is in the range 10–16 kg/h. Extrapolating from 80 W for 50 g/h to these flow rates, we obtain the power requirements for CO_2 acquisition and compression at full scale to be in the range 17–25 kW, depending primarily on crew size.

The average specific heat of CO_2 over the range from Martian temperature to 800 °C is 1 kJ/kg-K. To warm say 13 kg/h of CO_2 from 240 to 1100 K requires:

$$13 \times 860/3600 \, kJ/s \sim 3 \, kW$$

For oxygen-only ISRU via solid oxide electrolysis, the power requirement can be estimated by noting that the thermo-neutral voltage is about 1.46 V. If this voltage is multiplied by the current, the power for $CO_2 \Rightarrow O_2$ conversion is obtained. Since 1 g/h production of O_2 corresponds to 3.35 A, O_2 production at 2.3–3.3 kg/h requires 8–11 kW. In addition, heat losses are likely to require another 2–3 kW of power input. This does not include power for control and monitoring.

Over all, the power requirement for oxygen-only via SOXE with mechanical CO_2 acquisition is estimated to be roughly as follows.

Crew of four: 17 + 3 + 8 + 2 = 30 kW
Crew of six: 25 + 3 + 11 + 3 = 42 kW

Power to compress and condense oxygen and deliver to the MAV is not included.

The power requirement for a system that utilizes gypsum-rich regolith to extract water for a Sabatier process to produce CH_4 and O_2 depends on the factors shown in Table 1.14. For a system that produces 2.3–3.3 kg/h of O_2 (128–183 g-mol/h), 256–366 g-mol of water must be electrolyzed. It takes 237 kJ to electrolyze 1 g-mol of water. Therefore the power requirement is 61,000–87,000 kJ/h = 17–24 kW. The power requirements for the first three steps in Table 1.14 are unknown.

The power requirement for Mars ISRU (not including control, monitoring, and transfer of ISRU products to the MAV) depends on the crew size (4 or 6). For oxygen-only, the power requirement is in the range 30–42 kW. For water-based ISRU it is 34–49 kW, plus

Table 1.14 Power requirements for water-based Mars ISRU

Subsystem	Operational requirement	Power requirement (kW)
Excavator	Excavate regolith, deliver to water extractor, remove spent regolith	?
Water extractor	Heat regolith, condense water vapor, store water	?
Water transport	Deliver water from Extractor to Main ISRU process unit	?
Main processor	Acquire and compress CO_2	17–25
	Electrolyze water	17–24
	Produce $CH_4 + O_2$	Nil

the power needed to excavate regolith, deliver to water extractor, remove spent regolith. In general, it is expected that a 40 kW power system should suffice.

1.8.2.2   One Reactor Versus Several Smaller Ones

Several studies of power alternatives for Mars ISRU were conducted in the period 2015–2016 (e.g. Rucker et al. 2015, 2016b; Rucker 2016a; Palac et al. 2015). Assuming that a ∼40 kW power system might be needed, Rucker and co-workers studied the trade between using one fission space power system (FSP) of about 40 kW versus use of multiple smaller units such as four 10 kW fission space power systems. The authors made good points in support of using four 10-kWe units, although some issues remain.

A 40 kW fission space power system (FSP) has a roughly estimated mass of 7 mT and requires a radiator of roughly 184 m^2. That was based on use of a Stirling cycle engine to convert heat to power. (The efficiency was not stipulated but I suppose it might be 25%.) If a Stirling cycle power converter is not available, the radiator would undoubtedly have to be much larger. The FSP is delivered to Mars on a Cargo Lander with the Ascent Vehicle and ISRU system. In their model, it runs for 480 days to produce oxygen via solid-state electrolysis. However, as we have shown, it is likely it would run for only 420 days. Methane is delivered in the MAV. The FSP is deployed at least 1 km from the Cargo Lander with for radiation protection. The Cargo Lander must keep the methane liquefied and it fulfills other operational needs so it needs power from the time it arrives. As we stated, the FSP must be transported at least a km away, and a berm might be created for better radiation protection. A cable must be laid from the FSP to the Cargo Lander. The cable might possibly have a mass of ∼1 mT. Thus the Cargo Lander must have its own source of power for the interim set-up period that might last as much as ∼30–40 sols.

Approximately 26 months later, the crew lands with the Habitat, other equipment, and (as given in the original plan) a second 40 kWe FSP as backup for the previously delivered FSP. The crew cannot land within 1 km of the Cargo Lander for landing safety. This second FSP must also be at least 1 km from the Cargo Lander and the crew Habitat.

The mobility system required to deploy the FSP is regarded as a problematic challenge. The power system for the mobility system is not fully defined yet.

In the studies by Rucker et al., it was estimated that the power requirement for the first (ISRU) phase is 34 kW and the power requirement for the second (crew) phase is 33.5 kW. Hence the same FSP that supplied power for ISRU could then provide power to the crew when the MAV tanks are full and the crew lands.

Rucker et al. then explored the pros and cons of replacing the single 40 kW FSP with multiple smaller units, such as four 10 kW units or two 20 kW units. It was claimed that use of the smaller units allowed a mass saving due to a simpler method to transport heat to the Stirling engine (heat pipes replace pumped fluids). While the single 40 kW FSP had an estimated mass of 7 mT, the set of four 10 kWe had an estimated mass of 6 mT. (I can't imagine why the total area of the radiators was reduced from 184 to 80 m^2 since the conversion efficiency ought to be similar in both cases.) Deployment of four smaller units can be achieved with a much smaller mobility system than for use of one 40 kWe

FSP. However it seems likely that deploying four units and connecting them up might take more set-up time than a single unit.

Another point that was made is that the "spare" FSP probably need not include four 10 kWe units since it is very unlikely that all four FSP units would fail. If for example, only two 10 kWe units were included in the spare, that would save 3 mT.

It was claimed that use of four smaller FSP reduces the mass of the cable by a factor of ten. *How is this possible?*

The interim power system during set-up was not specified for the 40 kWe FSP case. However, it was claimed that with four 10 kWe units, one of these could be set up quickly near the Cargo Lander for interim power while the other three were deployed. *I would like to see more evidence that this is practical.*

Use of smaller FSP would make it much easier to deploy them, and move them to another site if that became useful. It was claimed that after a shutdown, the crew could approach a FSP after about a week. Movable FSP might allow exploration of a wider range of area on Mars. However deployment of four units with interconnections adds complexity. The single FSP might land with cables pre-connected.

Use of smaller FSP also supports sub-scale demonstrations. *However, a 7 mT FSP might be overkill for subscale ISRU demos.*

It was claimed that the stowed volume of four 10 kW FSP would be 22 m^3 versus 62 m^3 for a single 40 kWe FSP, using deployable radiators for the small unit and a fixed radiator for the large unit. But with fixed radiators, use of four smaller FSP requires a radiator of about 51 m^2. *It is not immediately clear why the larger unit cannot use deployable radiators.* The estimated areas of radiator will increase sharply if Stirling cycle engines are not available for power conversion.

All of the above estimates suggest that for FSP, there is no "economy of scale". *This seems counter-intuitive.*

Rucker et al. also listed a number of disadvantages attributable to use of four 10 kW FSP rather than one 40 kW FSP (e.g. deployment would require more rover trips, and cabling would be more complex).

1.8.2.3   Solar Versus Fission for Mars Surface Power

Rucker et al. compared solar to fission power at two levels of production:

- Precursor demonstration at 1/5 of full scale "ISRU Demonstrator" (0.45 kg/h O_2 production rate). Operation for 10,000 h, producing 4500 kg of O_2, with 1500 kg stored in a tank, and the balance vented. (*Based on crew of four.*)
- Full scale crewed mission (23 mT of O_2 in 420 days = 2.3 kg/h). (*As we have explained several times in this book, NASA continues to assume a 16-month period of ISRU production, while I claim the realistic period is 14 months.*) (*Based on crew of four.*)

ISRU Demonstrator:

Four options were examined for the ISRU demonstrator:

1. 6.5 kW fission power, using a nominal 10 kWe FSP operating at partial production rate.
2. Solar power with daylight-only ISRU operation (same O_2 production rate as fission) (Four 5.6 m dia arrays).
3. Solar power with daylight-only ISRU operation (double the production rate of fission —0.9 kg/h) (Four 7.5 m dia arrays)
4. Solar power with around-the-clock ISRU using batteries (0.45 kg/h).

The solar systems were located near the equator. It was assumed there would be one major dust storm lasting 120 sols. The solar conversion efficiency was optimistically assumed to be 33%. Li-Ion batteries were assumed with 60% depth of discharge, and 165 W/kg. The results of the study for the precursor mission are summarized in Table 1.15.

Table 1.15 Results of fission versus solar study

	Option solar 1A: 1/5 rate daytime only	Solar 1B: 1/5 rate around the clock	Solar 1C: 2/5 rate daytime only	Fission: 1/5 rate around the clock fission power
Total payload mass (including growth) (kg)	1128	2425	1531	2751
Electrical system mass (kg)	455	1733	639	1804
ISRU subsystem mass (kg)	192	192	335	192
Power	~8 kW Daylight	~8 kW Continuous (16 kW of arrays)	~16 kW Daylight	~7 kW Continuous
Solar arrays	4 each × 5.6 m diameter	4 each × 7.5 m dia.	4 each × 7.5 m diameter	None
Night production?	No	Yes	No	Yes
LOX production	4.5 kg/sol	10.8 kg/sol	9.0 kg/sol	10.8 kg/sol
Time to produce 4400 kg LOX, including 120 day dust storm outage	1098 sols	527 sols	609 sols	407 sols
ISRU on/off cycles	1098	<5	609	<5

Some observations made by the study:

- Daytime-only solar power concept offers the lowest landed mass but it requires 1100 sols with 1100 ISRU on/off cycles. (*This amount of cycling is probably a non-starter for any ISRU system.*)
- Fission power was at a comparative mass disadvantage in this trade because the 10 kW FSP was oversized for the 7 kW application, and the mass included a crew protection shield that wasn't necessary for demo.
- The equatorial site represents a minimum solar power mass; a significantly higher mass is expected at other latitudes. (*It is unclear whether gysum-rich sites can be found at the equator.*)
- It was claimed: "All options fit comfortably within allowable payload limits so mass alone is unlikely to drive a decision for an equatorial mission". (*But mass is only one of many considerations in choosing a power source. Furthermore, it is probably not a primary consideration. Reliability, lack of cycling, ease of set-up, and scalability might be more important.*)
- Power system selection probably depends on other factors. (*As stated above.*)
- Demonstrator mission solar power hardware costs are ∼ $100 M less than comparable fission power hardware costs. (*However, they did not provide the total cost of the demonstrator mission. Is $100 million a large or small percentage of the total cost? And as we stated above, reliability, lack of cycling, ease of set-up, and scalability would important as well. Furthermore, use of the nuclear option would open the door to further expanded use of nuclear power.*)

Full scale crewed mission:

Rucker et al. assumed up front that total need for oxygen production in a crewed surface mission was 23 mT, but this would pertain to a crew of four.

For ISRU to produce 22,728 kg of LOX in 420 days, the power needs were estimated as shown in Table 1.16. The estimated power required for ISRU appears to be low by a factor of about two.

Table 1.16 Power needs at full scale for a crew of four (Rucker et al.)

	Peak power needed (W)		Keep-alive power needed (W)	
Element	Cargo phase	Crew phase	Cargo phase	Crew phase
ISRU	19,700	0	19,700	0
MAV	6655	6655	6655	6655
Surface habitat	0	14,900	0	8000
Science laboratory	0	9544	0	174
Total	26,355	31,099	26,355	14,829

In the nuclear option, they used four 10 kW FSP, plus a fifth 10 kW FSP as backup. The total mass was estimated to be ~9 mT. (*If necessary, even three 10 kW FSP might suffice.*)

In the solar option, they would land solar arrays and energy storage with the first landing. This first delivery would provide about 35 kW day and night for ISRU, but during a dust storm, power would be reduced to about 11 kW. It is not clear whether sufficient propellants could be produced in this case.

Once the crew arrived, additional solar arrays would be deployed but no additional energy storage would be added. Surprisingly, power availability with extra arrays would only be 32 kW (day) and 27 kW (night). (*It does not make sense that additional arrays provide less power.*) With these additional arrays, loads drop to about 23 kW say and night during a dust storm. The total solar power mass is about 6 mT at the first delivery and rises to 11.7 mT with the second delivery. All of these estimates are for a favorable solar location.

Rucker (2016) reached the following conclusions:

Solar-powered crew surface mission is certainly possible, at least for some latitudes.

Solar power benefits from high technology readiness, lower cost, but it entails a high mass penalty, will limit landing site options, and has higher risk during a dust storm.

Fission is reliable, lower mass for most landing sites, provides the same mass regardless of site, season, day/night, or weather; but lower technology readiness and higher development cost.

Either power system will require substantial technology development and flight hardware investment.

The risks, uncertainties and variability of solar power remain unclear. For landing sites rich in gypsum, the case for solar might be much worse.

Lunar ISRU

2

2.1 Lunar Missions

NASA's Exploration Systems Architecture Study Final Report was on lunar mission design was published by NASA in 2005.[1] The lunar mission design section of this report[2] presented a historical review of lunar mission concepts, and used that as a basis for outlining the "ESAS" plan for a return of humans to the Moon. A "key finding" based on previous studies was given as:

> ISRU: Technologies for "living off the land" are needed to support a long-term strategy for human exploration. Key ISRU challenges include resource identification and characterization, excavation and extraction processes, consumable maintenance and usage capabilities, and advanced concepts for manufacturing other products from local resources.

A "functional requirement" was the ascent system must be "capable of utilizing locally produced propellants". At that time the preliminary mass allocation for a lunar landing included 600 kg for a "Lunar Miner/Hauler", 800 kg for an "ISRU O_2 Pilot Plant", 900 kg for an "Inchworm", 1000 kg for an "ISRU Logistics Carrier", and 1200 kg for a "ISRU Lunar Polar Resource Extractor". In discussing sites for the lunar out post, they emphasized sites with potential resources for ISRU. Any form of ISRU was considered fair game. For example, they mentioned a site good for processing "solar wind hydrogen", and other sites with "high-Ti pyroclastic glass may be excellent feedstock for ISRU processing".

Initial sortie missions would "conduct ISRU demonstrations, such as regolith excavation, manipulation, and processing".

---

[1] https://www.nasa.gov/exploration/news/ESAS_report.html
[2] https://www.nasa.gov/pdf/140635main_ESAS_04.pdf

© Springer International Publishing AG 2018
D. Rapp, *Use of Extraterrestrial Resources for Human Space Missions to Moon or Mars*, Springer Praxis Books, https://doi.org/10.1007/978-3-319-72694-6_2

This document provided the following paragraph:

During the outpost missions, the use of in situ resources will transition from demonstration to incorporation, and ISRU technologies successfully demonstrated on the sortie missions will be scaled up to production-level plants and facilities. Early activities will include … the production and storage of life support consumables such as oxygen for the crew's habitat and EVA systems will begin the transition from reliance on Earth-supplied logistics to self-sufficiency on the Moon. As production rates increase, lunar resources will provide the propellants needed by the landing spacecraft, which will lead to basing and servicing of reusable spacecraft on the Moon's surface.

New ISRU capabilities can be demonstrated and incorporated into mission plans on an as-needed and evolutionary basis to lower the cost of outpost missions and demonstrate capabilities that may be required for long-duration stays on the Mars surface. One area that may be critical for long-term lunar operations and trips to Mars where logistics management may be difficult is in situ manufacturing and repair. This capability includes the ability to fabricate spare parts—especially for high-wear excavation and regolith processing items— and repair techniques for both internal and external hardware. Fabrication processes of interest include additive, subtractive, and formative techniques for multiple feedstock materials (metals, plastics, and ceramics). A "machine shop" capability that includes repair techniques and part characterization may be required. Initially, feedstock from Earth can be used for manufacturing parts; however, resources from regolith to support this capability would be required for permanent surface operations.

The ability to extract metals (iron, titanium, aluminum, etc.) and silicon from regolith is of interest to support in situ manufacturing and construction capabilities that could be used to lower the cost of infrastructure growth during the outpost phase. Included in this work is development of other manufacturing and construction feedstock, such as concrete, wires, basaltic fibers and bars, metal tubing, etc. Several oxygen extraction concepts can be modified to include additional steps to extract these resources for use in construction feedstock.

Evidently, at the start of Griffin's Constellation Program, the ambitions for ISRU ran very high.

In the heyday of Griffin's Constellation Program:

NASA published their initial plan for a lunar exploration campaign… The campaign strategy was based on the establishment of an "outpost first" with continuous or near-continuous habitation capability on the rim of Shackleton Crater at the lunar South Pole; the outpost infrastructure is delivered on crewed flights in units of 6 mT size… This strategy presented in December 2006 is one of a large number of conceivable campaigns and associated lunar surface architectures. (Hofstetter et al. 2007)

Rapp (2015) provided a review of several mission models for a lunar outpost. If $CH_4 + O_2$ are used for ascent, it would take about 8.0 mT of oxygen (when combined with about 2.6 mT of methane) to ascend from the Moon, and with two ascents per year, the requirement is for $2 \times 10,600 = 21,200$ kg of ascent propellants at an $I_{sp} \sim 360$ s. If $H_2 + O_2$ is used for ascent the I_{sp} is ~ 450 s so the propellant need is $(360/450) \times 21,200 = 17,000$ kg. At a 1:6.5 ratio (H_2/O_2) the requirement is for $17,000/7.5 = 2260$ kg of H_2 for ascent per year (2 ascents). Some mission plans called for use of space storable propellants, in which case, ISRU would be moot.

With the advent of Project Constellation, funding for lunar ISRU increased substantially for several years. The output of these efforts is discussed in Sect. 2.2.

As Larson et al. (2010) said:

Over the past few years, the ISRU project has been focused on meeting the needs of the Constellation program. After several iterations of NASA study teams, the emerging consensus was that the primary role of ISRU for a Lunar Outpost would be the production of oxygen. The studies indicated that the Outpost would need approximately one metric ton of oxygen each year to close the life support loop.

Evidently, by 2010, NASA had switched from oxygen to space storable propellants for ascent, and the only (relatively minor) role remaining for ISRU was to produce a comparatively small amount of oxygen for life support.

2.2 Lunar Resources

The lunar regolith includes the atomic distributions shown in Table 2.1.

Lunar soil is generally fine-grained with over 95 wt% less than 1 mm; 50% is less than about 50 μm (the thickness of a human hair); and 10–20% is finer than 20 μm. The lunar-soil particle size distribution is very broad. In addition, because of the irregular particle shapes, the specific surface area is high: approximately 0.5 m^2/g. In fact, lunar soil particles have about eight times more surface area than an assemblage of spheres with the equivalent particle size distribution. As a result of both of these factors, lunar soil particles do not pack together as efficiently as, for example, uniform spheres. Even when lunar soil is packed extremely tightly (by a combination of compression and shaking!, the porosity is roughly 40–50%—high by terrestrial standards) (Taylor and Meek 2005).

According to McKay et al. (1992):

Lunar surface regolith is modified by two processes: (1) brecciation, including pulverization through meteoritic and cometary impact, and (2) solar wind implantation of volatile species. The longer the lunar regolith is exposed to bombardment, the greater the extent of pulverization and implantation of volatile species, However, two counterproductive processes are in operation. First, major impacts result in throw-out of large volumes of material that cover up older regolith. Second, micrometeorite bombardment results in production of agglutinates, that are lunar fines welded together with liquid silicate (glass). Thus, there are both destructive and constructive processes operating on the lunar surface.

The heating of the fines associated with micrometeoritic events liberates some of the solar-wind-implanted volatile species and plays an important role in the distribution and redistribution of volatiles on the Moon. The formation of agglutinates also decreases the availability of minerals for concentration from lunar fines (the particles that are less than 1 mm in diameter).

The concentration of imbedded solar wind atoms in the upper lunar regolith fines was measured to range from:

H: 10 to 120 ppm
C: 20 to 280 ppm
N: 20 to 160 ppm

Table 2.1 Average major element chemistry for mare and highland soil (McKay et al. 1992)

Element	Percent of atoms			Weight % of oxides		
	Mare	Highland	Average surface	Mare	Highland	Average surface
O	60.3	61.1	60.9			
Na	0.4	0.4	0.4	0.6	0.6	0.6
Mg	5.1	4.0	4.2	9.2	7.5	7.8
Al	6.5	10.1	9.4	14.9	24.0	22.2
Si	16.9	16.3	16.4	45.4	45.5	45.5
Ca	4.7	6.1	5.8	11.8	15.9	15.0
Ti	1.1	0.15	0.3	3.9	0.6	1.3
Fe	4.4	1.8	2.3	14.1	5.9	7.5

The higher concentrations are relatively rare.

According to Taylor and Meek (2005),

The major factor in the formation of lunar soil is micrometeorite impact. Rocks are broken into smaller pieces, accompanied by formation of considerable glass (typically 40–75%) from impact-produced melting. Two competing processes are operative: comminution of larger soil particles into smaller ones, and agglutination, where silicate melt welds together soil grains into glassy aggregates. These two competing processes complicate the formational characteristics of the soil. The lunar soil consists basically of the disaggregated pieces of the rock, but with a large component of impact-produced glass.

There are basically four potential lunar resources:

- Silicates in lunar highlands regolith containing typically >40% oxygen.
- Iron oxide in lunar Mares regolith containing on average 14% iron oxide, although some areas may have up to 25% iron oxide.
- Imbedded atoms in regolith from solar wind (typically several to many tens of parts per million).
- Water ice in regolith pores in permanently shadowed craters near the poles (unknown percentage but possibly a few percent in some locations—vertical and horizontal distributions are not known).

2.2.1   Silicates in Regolith

Lunar ISRU based on extraction of oxygen from lunar highlands regolith has two advantages:

(1) Regolith is typically >40% oxygen, which is a considerable amount.
(2) Regolith is available everywhere and solar energy may be applicable for processing.

Unfortunately, the oxygen in regolith is tied up in silicate bonds that are amongst the strongest chemical bonds known, and breaking these bonds inevitably requires very stringent conditions, particularly temperatures well above 1000 °C, and for some processes as high as 2500 °C. Some form of reducing agent is needed to extract the oxygen from silicates. This could typically be hydrogen or carbon.

Despite the great challenges involved in extracting oxygen from regolith, documents indicate that NASA has remained optimistic that they will succeed. But these technologies require reactors at very high temperatures that must take in lunar regolith and discharge spent regolith or slag. The probability that a practical process for autonomous lunar operation will result from this research appears to below.

In the unlikely case that a high-temperature processor for oxygen from regolith on the Moon can be made into a practical unit, one would still be faced with the challenges (and costs) for development and demonstration of autonomous ISRU systems for excavation of regolith, delivery of regolith to the high-temperature processor, operation of the high-temperature processor with free flow of regolith through it (with no caking, agglomeration and "gunking up" of regolith), and removal of spent regolith from the high-temperature processor to a waste dump.

2.2.2   FeO in Regolith

Ilmenite, olivine, pyroxene, and glass provide Fe-bearing phases in lunar soil in lunar Mare regions and these can be reduced by a hydrogen reduction process under much less stringent conditions than reduction of silicates. Ilmenite contains the highest concentration of FeO and has been found in lunar rocks in abundances above 25 wt% in some Mare regions. Beneficiation for the enrichment of ilmenite from the feedstock would help to decrease the amount of material handled in the hydrogen reduction process and therefore would reduce material processing requirements.

2.2.3   Imbedded Atoms in Regolith from Solar Wind

An Internet site provided this paragraph[3]:

> Because the Moon has no atmosphere and the solar wind ions are moving fast, they are imbedded or implanted into the surface material of the Moon. They do not penetrate very deep into a rock or mineral grain, only a few hundredths of a micrometer, so all the solar-wind-implanted atoms are at the very surface of lunar regolith grains. Meteorite impacts stir the surface regolith so that the upper few meters of regolith are rich in implanted ions of hydrogen and helium. The amount of solar wind implanted ions is greater in the very finest material because the fine material has more surface area than the coarser material.

---

[3]http://meteorites.wustl.edu/lunar/regolith_breccia.htm

2.2.4   Water Ice in Regolith Pores in Permanently Shadowed Craters Near the Poles

The following is abstracted from a NASA website[4]:

The Lunar Prospector spacecraft indicated that water ice might be present at both the north and south lunar poles, in agreement with interpretations of Clementine results for the South Pole reported in 1996. The ice originally appeared to be mixed in with the lunar regolith (surface rocks, soil, and dust) at low concentrations conservatively estimated at 0.3–1%. Subsequent data by the Neutron Spectrometer on the Lunar Prospector spacecraft taken over a longer period has indicated the possible presence of discrete, confined, water ice deposits buried beneath as much as 40 cm of dry regolith, with the water signature being stronger at the Moon's north pole than at the south. The ice is thought to be spread over roughly 1850 km^2 at each pole. The total mass of ice could be as high as a few billion mT.

The Moon has no atmosphere so that any substance on the lunar surface is exposed directly to a vacuum. This implies that water ice will rapidly sublime directly into water vapor and escape into space, because the Moon's low gravity cannot hold gas for any appreciable time. Over the course of a lunar day ($\sim$29 Earth days), all regions of the Moon are exposed to sunlight, and the temperature on the Moon in direct sunlight reaches as high as 400 K, so any ice exposed to sunlight for even a short time would be lost. The only possible way for ice to exist on the Moon would be in a permanently shadowed area within a polar crater.

The Clementine imaging experiment showed that such permanently shadowed areas do exist in the bottom of deep craters near the Moon's south pole. In fact, it appears that approximately 6000–15,000 km^2 of area around the South Pole are permanently shadowed. The permanently shadowed area near the North Pole appears on Clementine images to be considerably less extensive but the Lunar Prospector results show a much larger water-bearing area at the North Pole. Much of the area around the South Pole is within the South Pole-Aitken Basin, a giant impact crater 2500 km in diameter and 12 km deep at its lowest point. Many smaller craters exist on the floor of this basin. Since they are down in this basin, the floors of many of these craters are never exposed to sunlight. Within these craters the temperatures would never rise above about 100 K. Any water ice at the bottom of the crater could probably exist for billions of years at these temperatures.

The Moon's surface is continuously bombarded by meteorites and micrometeorites. Many, if not most, of these impactors contain water ice, and the lunar craters show that many of these were very large objects. Any ice that survived impact would be scattered over the lunar surface. Most of this ice would be quickly vaporized by sunlight and lost to space, but some would end up inside the permanently shadowed craters, either by directly entering the crater or migrating over the surface as randomly moving individual molecules that would reach the craters and freeze there. Once inside the crater, the ice would be

---

[4]http://nssdc.gsfc.nasa.gov/planetary/ice/ice_moon.html

relatively stable, so over time, the ice would collect in these "cold traps", and be buried to some extent by "meteoritic gardening".

In 2009, the impact plumes of the Lunar Crater Observation and Sensing Satellite (LCROSS) and its Centaur rocket stage in Cabeus crater near the south pole of the Moon showed the spectral signature of hydroxyl, a key indicator that water ice is present in the floor of the crater.

2.3   Lunar ISRU Processes

In 2006, NASA carried out an exercise in which they summarized the quantities involved in processing lunar resources for the then evolving plans to return to the Moon. I was fortunate to gain access to this work, which is not available to the general public. These studies were tabulated in brief reports called "baseball cards" although it is not clear to this writer why this connection was made. In the following sections, I review the results of the baseball cards with commentary.

2.3.1   Oxygen from FeO in Regolith

The hydrogen reduction system operates on the principle of reducing metal oxides, mainly iron oxide and its derivatives within the lunar regolith. This system is conceived as operating continuously during the lunar day (two and a half weeks) while solar energy is available. Heated regolith feedstock reacts with hydrogen (supplied from Earth) to produce water. The reaction temperature for this process is in the range of 1200–1300 K. Product water is then electrolyzed to regenerate reactant hydrogen and liberate oxygen. For a hydrogen reduction system, the process chemistry is not very complicated and re-supply masses for reagent makeup due to process losses are expected to be minor. Process temperatures are below the melting point of the regolith feed. Ilmenite, olivine, pyroxene, and glass are the dominant Fe-bearing phases in lunar soil these can all be reduced by this hydrogen reduction process. Ilmenite contains the highest concentration of FeO and has been found in lunar rocks in abundances above 25 wt% in Mare regions. However, iron concentrations in Highlands areas have much less iron content, with typically only 5% FeO by weight. Oxygen yields were claimed to be only 3% for these soils. *Actually, the maximum oxygen yield is $(16/72) \times 5\% = 1.1\%$ assuming 100% recovery, where $16/72 = O/(O + Fe)$*. Beneficiation for the enrichment of ilmenite from the feedstock would help to decrease the amount of material handled in the hydrogen reduction process and therefore would reduce mass processing requirements. Reaction times of one hour were anticipated per batch.

The oxygen production process begins when regolith is transported from a hopper to the reactor by an auger system. The auger system additionally functions as a heat exchanger to recuperate heat from the hot regolith being removed from the reactor. This concept requires the use of three reactor beds, one of which is for preheating, one is for emptying and one is for reacting—at all times. *However, a remote, autonomous heat exchanger in which the two counter-flow materials consist of regolith might be subject to many maladies, particularly the "true grit" might be subject to "gunking up".*

The reactor assembly is a multi-reactor system, where the reactors are operated in alternating mode in order to maximize reaction time and to reduce power consumption by increasing the heating time. Water produced within the reactor is directed to a water electrolyzer, where the water is split into oxygen and hydrogen. The hydrogen is recirculated back to the reactor and the water is re-circulated to the electrolyzer. Oxygen produced in the electrolyzer is delivered to the oxygen tanker for liquefaction and storage.

Summary of Concept of Operation:

1. The following systems are delivered to the surface of the Moon: excavator/hauler, oxygen production plant, liquid oxygen tanker, and surface cryo depot.
2. The excavator/hauler collects and transport regolith from the excavation site to the oxygen production plant and dumps it into the hopper.
3. The excavator/hauler is loaded with spent regolith and transported to the dumping area.
4. Regolith is moved simultaneously from the hopper to the reactor and from the reactor to the hauler via a countercurrent heat exchanger auger to recuperate an assumed 50% of the heat.
5. Hydrogen flow begins when the regolith has reached reaction temperature (1200 K).
6. The water produced in the reactor is condensed, purified, and sent to a proton exchange membrane (PEM) electrolyzer.
7. Wet hydrogen out of the electrolyzer is sent to a condenser to condense the water. Condensed water is re-circulated to the electrolyzer and the hydrogen is re-circulated to the reactor.
8. Wet oxygen is sent to a condenser followed by a dryer to reduce the water content to 25 ppm.
9. Dry-gaseous oxygen is sent to the surface cryo depot for liquefaction and storage.
10. Stored liquid oxygen in the surface cryo depot is delivered to end-users by the liquid oxygen tanker.

Initial modeling exercises for predicting the overall system mass and power requirements for various oxygen production mass rates using hydrogen reduction were developed by NASA and presented in baseball cards as shown in Tables 2.2 and 2.3.

Table 2.3 is essentially a duplicate of Table 2.2 except for one difference: the iron oxide concentration is assumed to be 5% for highlands regolith instead of 14% for Mare regolith. This means that all the figures for the amount of regolith processed, and therefore the energy to heat the regolith from Mare regolith must be multiplied by $(14/5) = 2.8$ to get equivalent values for highlands regolith.

Table 2.2 Processing system attributes for hydrogen reduction of FeO in Mare regolith according to baseball cards

Annual oxygen production rate	1 mT	10 mT	50 mT	100 mT
System mass (kg)	88	469	1960	3919
Contingency mass (kg)	18	94	392	784
Total mass (kg)	106	563	2352	4703
System volume (m^3)	1.9	19	28	55
System power (kW)	1.4	13	61	122
Regolith feedstock (kg/h)	9	83	200	400

Table 2.3 Processing system attributes for hydrogen reduction of FeO in highlands regolith according to baseball cards

Annual oxygen production rate	1 mT	10 mT	50 mT	100 mT
System mass (kg)	88	785	2773	5546
Contingency mass (kg)	18	157	555	1109
Total mass (kg)	106	942	3328	6655
System volume (m^3)	1.9	21	26	51
System power (kW)	1.1	10	47	94
Regolith feedstock (kg/h)	16	183	827	1654

I have independently estimated the power requirements as follows:

The need is to heat X kg of regolith from 200 to 1300 K. The claim is made to recuperate 50% of heat in the heat exchanger and the heat loss from the reactor is estimated at 10%. It is difficult to substantiate these claims that seem optimistic, but they will be accepted for this illustration. The required heat is:

Heat = (X kg) (0.00023 kWh/kg K) (1100 K) (0.5 + 0.1) ≈ (0.15 X) kWh.

The regolith mass requirements were not given. However, we can estimate that. It is stated that Mare regolith is 14% iron oxide by weight, which implies that the oxygen content (of the iron oxide) constitutes about $0.22 \times 14\% = 3.1\%$ of the regolith mass. If we further assume that the oxygen extraction process is 90% efficient at removing this oxygen, we conclude that the required regolith mass is the mass of oxygen produced divided by (0.031×0.9) or $M(O_2)/(0.028)$. The results are shown in Table 2.4. These results for power requirements are close to those given in Table 2.3.

As we pointed out previously, the power requirement for processing FeO in highlands regolith should be about a factor of 2.8 times the power requirement for processing FeO in Mare regolith. The result is shown in Table 2.5. The power requirements are much higher than those given in Table 2.4.

Table 2.4 Power requirements for processing FeO in Mare regolith

Annual oxygen production rate	1 mT	10 mT	50 mT	100 mT
Annual regolith rate (mT)	33.6	336	1680	3361
Annual regolith rate (kg)	33,613	336,134	1,680,672	3,361,344
kWh	5042	50,420	252,100	504,201
Hours	3500	3500	3500	3500
kW to heat regolith	1.44	14.4	72	144

Table 2.5 Power requirements for processing FeO in highlands regolith

Annual oxygen production rate	1 mT	10 mT	50 mT	100 mT
Annual regolith rate (mT)	94	941	4706	9412
annual regolith rate (kg)	94,118	941,176	4,705,882	9,411,765
kWh	14,118	141,176	705,882	1,411,765
Hours	3500	3500	3500	3500
kW to heat regolith	4.0	40	202	403

2.3.2   Oxygen Production from Silicates in Regolith

This concept is based upon a high-temperature, direct energy processing technique to produce oxygen, silicon, iron, and ceramic materials from lunar regolith via carbonaceous high-temperature (carbothermal) reduction. The direct heating source only processes a localized region of regolith at any given moment. The surrounding regolith acts as an insulating barrier to protect the support structures. Concentrated solar energy is beamed down onto a central portion of the top of the crucible filled with regolith.

The percent oxygen recovered per mT of regolith processed is substantially higher than for hydrogen reduction. This allows the process to be used in locations with little iron content, such as highlands locations, as well as Mare locations. The high reaction temperatures (up to 2600 K) are achieved in a small spot of highly concentrated sunlight. Initial methane and hydrogen feedstocks are regenerated via a catalytic methanation reactor and a water electrolyzer. System inputs are regolith, highly concentrated solar heat energy, and DC electric power to electrolyze water produced in the process. Oxygen and reduced slag are the outputs.

The concept of operation is as follows: Screened and crushed regolith is placed on a circular array of reactor cells. A solar concentrator is used to provide direct heating of the regolith at the surface in the carbothermal reduction cells. Methane gas is injected into the reduction chamber. The lunar regolith absorbs the solar energy and forms a small region of molten regolith at the surface in the center. A layer of unmelted regolith underneath the molten region insulates the processing tray from the molten region. Methane gas in the reduction chamber cracks on the surface of the molten regolith producing carbon and hydrogen. The carbon diffuses into the molten regolith and reduces the oxides in the melt while the hydrogen gas is released into the chamber. A moveable solar concentrator allows heating in the form of a highly concentrated beam focused on the regolith surface. A system of fiber optic cables distributes the concentrated solar power to small cavities formed by reflector cups that concentrate and refocus any reflected energy. Solidified slag melts are removed from the regolith bed by a rake system. Slag waste and incoming fresh regolith are moved out or into the chamber through a double airlock system to minimize the loss of reactive gases.

A short time after the solar heating process begins the product gases from the reduction chamber are slowly pumped through the catalytic reactor bed. The ceramic glass material would fall into a collection bin at the bottom of the reduction chamber. A sealed hopper assembly located on the cover is used to meter fresh regolith into the reduction chamber. The fresh regolith is evenly spread over the tray by an arm. The reduction chamber has gas ports to allow methane gas to be injected into the chamber and product gases to be pumped out of the chamber.

The catalytic reactor bed converts the carbon monoxide and hydrogen gases produced in the carbothermal reduction chamber into water and methane gas. A small hydrogen gas reserve is needed during the initial start-up process. However, no additional hydrogen is required once the carbothermal process reaches steady-state operation (i.e. when water is being electrolyzed to produce the hydrogen gas required).

Water must be separated from the mixture of hot gases that exit the catalyst bed reactor. The baseline design for the water separation unit is a simple water condenser. The reformed methane gas passes down through the collection reservoir and out to the carbothermal reduction chamber to reduce more regolith.

The liquid water that is collected in the water reservoir is fed into a PEM (Proton Exchange Membrane) electrolysis unit to produce oxygen and hydrogen gas. The electrolyzer system incorporates a re-circulating water system to keep the electrolyzer cool and drop out excess water from the outgoing hydrogen stream before it enters the catalytic reactor. The oxygen gas will also pass through a chiller and dryer before it is passed-on to the oxygen storage and liquefaction system.

Summary of Concept of Operation:

1. The following systems are delivered to the surface of the moon: excavator/hauler, oxygen production plant, liquid oxygen tanker, and surface cryo depot.
2. The excavator/hauler collects and transport regolith from the excavation site to the oxygen production plant and dumps it into the hopper.
3. The excavator/hauler is loaded with spent regolith and transported to the dumping area.
4. Solar concentrator(s) direct solar energy to the reaction chambers and methane flow is initiated.
5. Carbon monoxide produced is sent to a gas purification unit to remove sulfur gas.

6. Carbon monoxide is reacted with hydrogen in a catalytic reactor to produced water and methane.
7. The water produced in the reactor is condensed, purified, and sent to a proton exchange membrane (PEM) electrolyzer.
8. Methane is re-circulated to the regolith reduction reactor.
9. Wet hydrogen produced in the electrolyzer is sent to a condenser to condense the water. Condensed water is re-circulated to the electrolyzer and the hydrogen is re-circulated to the catalytic reactor.
10. Wet oxygen is sent to a water removal unit to reduce the water content to 25 ppm.
11. Dry-gaseous oxygen is sent to the surface cryogenic depot for liquefaction and storage.
12. Stored liquid oxygen in the surface cryogenic depot is delivered to end-users by the liquid oxygen tanker.

This process is extremely challenging from materials, thermal, optical and chemical points of view. Rube Goldberg should be reincarnated to see this one. It seems to be eminently complex, impractical and problematic, and there is no evidence that it could operate autonomously. Only a visionary with no practical experience could come up with this incredible scheme. This concept is so outlandish that it is not worth the effort to estimate the mass/power requirements.

2.3.3   Volatiles from Imbedded Atoms in Regolith from Solar Wind

Basic Quantities:

Analysis of lunar rocks from the Apollo missions indicated that heating of the lunar rocks evolved a variety of volatile materials. In 2006, NASA "addressed the prospect of extracting hydrogen for use as a propellant and nitrogen for use as breathing air and habitat conditioning". The assumed annual quantities of H_2 and N_2 required were 7775 kg/year (say ~ 8000 kg) for propellant for two annual launches from the Moon and 1500 kg/year for breathing air and habitat. For purposes of designing volatile extraction equipment, a best case (and unsubstantiated) assumption was that the concentrations of both H_2 and N_2 are at 150 ppm, that all of this embedded H_2 and N_2 can be extracted during the heating process, that no H_2 or N_2 is lost during the extraction processing, and that all of the H_2 and N_2 are collected and stored pressurized to 600 psi in tanks.

As we pointed out in Sect. 2.2.1, it takes about 8000 kg of oxygen (when combined with about 2600 kg of methane) to ascend from the Moon, and with two ascents per year, the requirement is for $2 \times 10,600 = 21,200$ kg of ascent propellants at an $I_{sp} \sim 360$ s. *There is no way that 7775 kg of H_2 is needed for ascent.*

The stated requirement for N_2 for breathing air at 1500 kg/year for breathing air and habitat is a little off. The rule of thumb is breathing oxygen is 1 kg/day per crewmember or $4 \times 365 \times 1 = 1460$ kg of O_2 per year for a crew of 4. Another rule of thumb is to use

1 part O_2 to 3 parts N_2 so the breathing N_2 requirement amounts to $3 \times 1460 = 4380$ kg per year. Thus, it is the N_2, not the H_2 that sets the size of the system. Yet the baseball card claimed: "Since more hydrogen is needed than nitrogen, the processing rate here is based on the hydrogen requirement."

The baseball card went on to decide on a processing method of volatile extraction. A primary driver for deciding on which process to employ is to quantify the amount of material that needs to be processed. Since more hydrogen is claimed to be needed than nitrogen, the processing rate was based on the hydrogen requirement. At a hydrogen concentration of 150 ppm, 53,333,333 kg/year of regolith would be processed in order to produce 8000 kg/year of H_2. Assuming that the regolith has a bulk density of 1500 kg/m^3, the volumetric processing rate is 35,556 m^3/year or about 5.8 m^3/h assuming that operations could be carried out for 70% of the hours in a year (Table 2.6).

As we pointed out in the previous paragraph, the proper result is that the requirement is driven by the need for 4380 kg per year of N_2 and the requirement for H_2 is only 2260 kg/year. Furthermore, it seems more appropriate to be a bit less optimistic about average concentrations and percent of hours of operation. I would use 75 ppm and 40% of hours for operation (Table 2.7).

The baseball card did not evaluate the power requirement for processing volatiles from regolith. One must process 5.6 m^3/h or roughly 8400 kg/h of regolith. If we assume that the specific heat of the regolith is 0.2 that of water, and assume that one must raise the temperature of the regolith by 600 °C to drive off volatiles, the power requirement is (8400 kg/h) (0.00008 kWh/kg K) (600 K) = 400 kWh per hour or a steady rate of 400 kW just for heating the regolith. While this power requirement can be reduced somewhat by using regenerative heating whereby hot regolith exiting the reactor heats incoming cold regolith, such a heat exchanger might be difficult to implement. Clearly, the power requirement is far too high to be feasible (Table 2.7).

Processing Scenarios:

The baseball card presented several concepts for batch processing along with one process that is semi-continuous. They rejected a batch process where a quantity of harvested regolith is periodically inserted into one of a series of batch heater processors each hour. They also rejected a second batch process concept in which a dome encloses an auger system, the auger digs a hole in the regolith within the enclosed regolith area, and the regolith is heated as it is excavated by an auger to evolve embedded volatiles.

Table 2.6 Regolith needed for volatile extraction according to baseball card

	Units	H$_2$ (8000 kg/year)	N$_2$ (1500 kg/year)
Regolith to be processed	m^3/year	35,600	6670
Regolith to be processed	kg/year	53,333,300	10,000,000
Regolith processed @ 70% of time	m^3/h	5.80	1.09
Regolith processed @ 70% of time	kg/h	8700	1630

Table 2.7 Regolith needed for volatile extraction according to this book

	Units	H$_2$ (2260 kg/year)	N$_2$ (4380 kg/year)
Regolith to be processed	m^3/year	10,000	19,470
Regolith to be processed	kg/year	15,000,000	29,200,000
Regolith processed @ 40% of time	m^3/h	2.8	5.57
Regolith processed @ 40% of time	kg/h	4300	8330

They selected a third batch-processing scenario for further consideration. In this process, there is a large inflatable dome that has a center-driven scraper-wand similar to an agricultural silo top-unloading device. The scraper moves in a circular sweep at a fairly slow rate, scraping a thin layer (say ~1 cm) of the surface layer per pass. The regolith that is scraped from the surface moves radially outboard until near the outer diameter it flows continuously over a weir onto a conveyor that is attached to and travels with the circular sweeper. The conveyor moves the regolith up a ramp where the scraped regolith is ultimately dumped to the outer periphery of the inside of the dome and improves the seal of the dome as processing proceeds. As the regolith moves up the conveyor it is heated (presumably by either microwave or IR heaters) to evolve embedded volatiles. The evolved N$_2$ is captured with a cryocooler and the evolved H$_2$ is captured by a metal hydride bed. Spent, processed regolith is dropped circumferentially along the inside diameter of the dome to provide a better and better seal as the processing progresses.

They considered two different sizes for producing 2000 kg/year hydrogen: (1) one 8.5 m diameter dome with a 2.5 m sweep wand radius; and (2) three 5 m diameter domes with 1.75 m sweep wand radii. In either case, the dome system is an inflatable rigidizable structure with reinforced strength at the location where the dome bottom circumference meets the regolith surface. The conveyor, the sweeper wand, and their drive motors must be constructed in place after inflation of the dome(s). Further, the domes would be mounted on a four-wheel drive system, such that once processing at a particular site has been completed, the dome can be hydraulically jacked up and then driven to the next adjacent processing site where it is jacked down again to interface with the surface for a new run. Movement of the domes would be required once every 24 h for these design sizes.

For the larger size, the wand is swept at a rate of 0.134 rpm and scrapes at a depth of 1 cm for each revolution. For a 24-h operation, 35 m^3 of regolith is processed over a 22-h period. The remaining 2 h are to be used to move the dome and equipment to a new location and get ready for the subsequent day's run. The regolith is removed at a diameter of 5 m to a depth of 1.8 m. After 22 h of operation, processing is stopped, during the following 2 h the dome along with it's ancillary equipment is raised and driven to a new adjacent processing location and the process repeated during the next 24 h period.

Similarly, for the smaller size (5 m in diameter), the wand is swept at a rate of 0.115 rpm and scrapes at a depth of 1 cm for each revolution. For a 24-h operation, 11.7 m^3 of regolith is processed in each of three domes for a total regolith processed in that 24-h period of 3 times 11.7 or 35 m^3. The regolith is removed at a diameter of 3.5 m to a depth of 1.2 m. After 24 h of operation, processing is stopped, the dome along with its ancillary equipment is raised and driven to a new adjacent processing location, and the process is repeated during the next 24 h period.

Concept Assumptions

1. An inflatable rigidizable dome can be built that can be deployed on the lunar surface and moved from place to place, yet is robust enough that it stays intact without leakage to the lunar environment for multiple years.
2. The sizing is based on production of 2000 kg/year of H_2.
3. All embedded H_2 is evolved, all evolved H_2 is captured and recovered and none is lost.
4. The equipment can be operated for 70% of the overall hours in any 365 day period.
5. The scraper will be able to scrape a 1 cm layer per revolution using lunar gravity only. Spring loading of the sweep wands is not required.
6. The scraper wand assembly is extractable from the dig hole after every processing period with no trouble.
7. The dome is movable from location to location on a four-wheel drive system.
8. The heat transfer rate into the regolith is high enough to heat the regolith from 0 to 500 °C by the time the regolith is conveyed to the inner periphery of the dome.

It is too bad that Rube Goldberg is not alive to enjoy this crazy, impractical scheme. This process is so outlandish that it is not worth the effort to quantify.

2.3.4 Water Extraction from Regolith Pores in Permanently Shadowed Craters Near the Poles

Concept Overview

The baseball card concept for water extraction from regolith pores in permanently shadowed craters near the poles is illustrated in Fig. 2.1. Excavation of ice from the regolith takes place some 8 km within a permanently shadowed crater. Regolith bearing water ice is transported from the excavation site about 100 m to a water extraction plant where the regolith is heated to sublime the ice to vapor. This water vapor is driven from the reaction chamber and passes over a high surface area, radiative freezer driven by the pressure differential between the reaction chamber and the lunar vacuum. As this freezer becomes fully loaded, it is isolated from vacuum of space and the reaction chamber is heated, melting the ice into liquid. The liquid is gathered in a sump and then pumped to a storage tank. When full, this water storage tank is transported some 8 km across the crater, and then up the steep boundary of the crater a considerable height, to the crater rim where an electrolysis system is placed for further conversion. The electrolysis unit requires the greatest amount of energy so it is placed on the crater rim where it is claimed that solar energy is available 70% of the year with outages less than 100 h at a time.

It is difficult to provide power within the crater. NASA has indicated:

> At the lower production rates the water extraction plant is powered by RTGs and at higher production rates by rechargeable battery sets that are charged in the light at the electrolysis unit and exchanged as the water tanks are exchanged.

However, the power requirements within the crater are far too great to be supplied by RTGs, and there is not enough plutonium available to fuel the large number of RTGs that would be required. In addition, RTGs are notoriously expensive. Similarly, the proposed battery packs would be enormous and might have difficulty functioning in a 100 K environment. Others have proposed power beaming from the upper crater rim where solar energy is plentiful but this seems to be another Rube Goldberg idea. It is not clear how the water would be transported over 8 km of crater surface. Would it be driven by astronauts, or transferred by robotic rover? How long would the traverse take? If astronauts are required, what is the cost of maintaining them? And can they stand exposure to RTGs?

Concept of Operations

According to the baseball cards, the water extraction plant is landed (with an excavator/hauler) in a permanently shadowed crater. Mechanical elements are deployed for operation. The excavator/hauler begins excavating regolith near the water extraction plant. The upper part of the regolith is desiccated, thus considerable overburden must be removed to make the water-bearing regolith accessible to the excavator/hauler. Comment: Ice at very low temperatures ($\sim$100 K or lower) becomes very hard. It might require heavy equipment or even explosives to break it up. Power availability is a major problem here as discussed above. The excavator/hauler delivers loads of regolith to the input hopper and it is then conveyed to the reactor by an auger for batch heating. When loaded with a full charge, the reactor is sealed and the regolith is heated with internal resistance heaters from approximately 100 to 270 K. According to NASA, this volatilizes the water in the regolith and drives the pressure up in the reactor causing the water vapor to flow out of the chamber to an orifice open to lunar vacuum. Comment: It is not clear whether the water ice is free ice or whether it is absorbed into a regolith matrix. Bringing the temperature up to 270 K might produce slow sublimation under these conditions. However, it seems likely that it might be necessary to heat the regolith/ice mixture to perhaps something like 373 K to boil off water at 1 bar.

In the NASA plan, the outgoing flow of water vapor is routed through a high surface area freezer that is cooled by a radiator with a view of deep space. The freezer surface area is sized such that the water liberated in a single charge can be frozen in a single cycle. After all the water from a charge has been frozen, the freezer section is isolated from both lunar vacuum and the reactor with valves. The freezer is then heated causing the ice to melt and drain by gravity to the bottom of a sump in the freezer chamber. At this point the water is pumped into a storage tank. When the water storage tank is full, it is removed

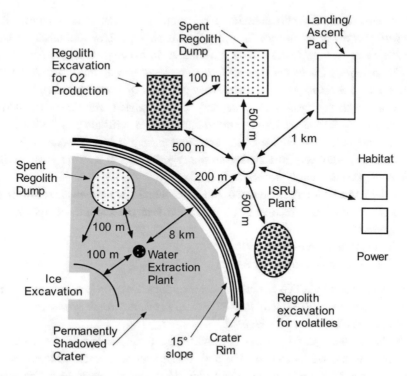

Fig. 2.1 Layout of NASA Lunar Polar outpost

(and replaced with an empty tank) by a tank transporter that travels about 8 km of crater surface and then up a ~15° slope to the crater rim to deliver the water tank to the electrolysis unit that converts the water into oxygen and hydrogen for use. *As stated previously, power availability within the crater is a major problem. If humans are required to drive the water tank 8 km across the crater and then up the crater side, the cost of maintaining these humans should be added to the cost of ISRU. Depending on how long it takes to traverse the 8 km, the required number of water tanks and delivery rovers might be large.*

According to NASA, concurrent with the step whereby the frozen water is warmed, the now dry regolith is removed from the reactor via the same auger system that delivered regolith to the heating chamber. The auger system additionally functions as a heat exchanger to recuperate the heat from the hot regolith being removed from the reactor. *This step could be problematic depending on the tendency of wet or dry regolith to agglomerate and gunk up.*

NASA sized the water extraction plant for 4 oxygen production rates. They made the following assumptions:

- The upper 10 cm of regolith is desiccated. *However, measurements with neutron spectrometer suggest 40 cm but this is unknown.*

- The water-bearing regolith contains 1.5% water by mass. *It is not clear how the percentage of water ice in regolith affects its morphology. If orbital measurements find that the water content is 1.5% over a 50 km × 50 km region, surely it is not unreasonable to expect (or at least hope) that some local areas within that footprint have more than 1.5%? This further exacerbates the issue of "consolidation."*
- The water-bearing regolith is loose and unconsolidated. *As the water percentage approaches 10–20% it will tend to consolidate into a rigid mass. Whether it remains "unconsolidated" at 1.5% water ice, is uncertain, but it is a reasonable hope.*
- The water is completely liberated from the regolith at 270 K. *This seems unlikely. A higher temperature might be needed.*
- The water extraction unit is operated 700 h per month. *See comment below on timing.*
- The dual auger system recuperates 70% of the heat per regolith charge. *This seems very optimistic.*
- Heat losses to the environment are 10%.
- Water in the tank is allowed to freeze.

Power is provided by RTGs for lower power options. *See previous comments on power within the crater. There is no way that RTGs can provide enough power and there is no way that enough RTGs can be obtained.*

For higher power options, power is provided by battery sets that are charged at the electrolysis unit and exchanged by the tank transporter with each tank exchange. *See previous comments on power within the crater. This does not seem to be feasible.*

Control is completely autonomous with periodic reporting of status to the ground. *This seems to imply that the entire process is autonomous without human intervention. There are so many things that could go wrong that this process seems to require human oversight.*

It must be assumed that excavation and hauling regolith to the water extraction unit is rapid and the 3 allocated haulers can continuously supply the water extraction unit with fresh regolith as fast as it is needed. It is also assumed that 3 additional haulers are needed to haul desiccated regolith to a dump.

Comment on timing: Let us denote the following durations for various steps:

Time to acquire water ice from a batch of regolith = t_1
Number of batches of regolith needed to fill water storage tank = N_1
Time to fill water storage tank = $N_1 \times t_1$
Time required to transport storage tank to crater rim = t_2
Time for round trip of storage tank to and from crater rim = $2 \times t_2$

Unfortunately, NASA did not provide an estimate of these timings. It seems likely to this writer that $t_2 \gg t_1$. In order to assure that there is always a water tank available to receive water from the water extraction unit, there must be several water tankers to haul water to and from the crater rim. If a water tanker gets filled in time t_1, the previous tanker won't return until $(2 \times t_2 - t_1)$ later. Hence the number of water tankers required is

$(2 \times t_2 - t_1)/t_1$. A wild guess is that t_1 might be of the order of 2 h. It is not clear how fast a rover carrying water can traverse the 8 km to the crater rim, climb up the slope, and deposit its frozen water in the electrolysis processing station. I would make a wild guess that t_2 might be something like 20 h. In that case, about 20 water tankers are needed, each equipped with its own power source. At any given moment, these tankers are arrayed so that 10 tankers are lined up every 0.8 km on a path going to the crater rim, and 10 tankers are lined up on a parallel path coming from the crater rim (see Fig. 2.2). In order to reduce the number of water tankers, each tanker could be made larger, but then the loads would be higher and the power requirements would increase per tanker. *Evidently, the whole scheme is completely impractical.*

Power is required within the crater for excavation and hauling regolith, for water extraction from regolith, for the water tankers, and for processing at the crater rim. First consider the water extraction unit. NASA provided the data shown in Table 2.8.

The data in Table 2.8 are very optimistic. My rendition is as follows. At 1.5% water content it takes 67 kg of regolith to yield 1 kg of water. But allowing for a desiccated layer, it may require at least 80 kg of regolith to produce 1 kg of water (probably more). 1 kg of water can generate roughly 1 kg of oxygen. I assume that the regolith must be heated from 40 to 380 K for a ΔT of 340 K. Table 2.9 shows the results for power requirements that are considerably higher levels than those estimated by NASA.

NASA estimated mass and power requirements for all elements of the scheme to extract water from permanently shadowed craters. Even their optimistic estimates are very high. For example, they estimated the power requirement for excavation and hauling of regolith in the crater to produce 10 mT/year of oxygen to be 66 kW. *Exactly how they would generate 66 kW deep in the crater is beyond my imagination.*

2.4 NASA Accomplishments and Plans

Prior to the NASA decision to return humans to the Moon made in 2004–5, ISRU technology was funded by NASA in dribs and drabs and only fledgling progress was made. With the advent of Griffin's lunar initiative, "NASA, under the Exploration Technology Development Program made significant investments in the development of Space resource utilization technologies as a part of the In Situ Resource Utilization (ISRU) project" (Larson et al. 2010)—to be denoted "LSS10". Reporting on their work over the time period 2006–2010, these authors claimed "the ISRU project has taken what was essentially an academic topic with lots of experimentation but little engineering and produced near-full-scale systems that have been demonstrated". They were referring to "systems that could produce oxygen from lunar soils". This paper reviewed past accomplishments of the NASA ISRU project and discussed plans for the project's future "as NASA moves to explore a new paradigm for space exploration that includes orbital fuel depots and even refueling on other planetary bodies in the solar system".

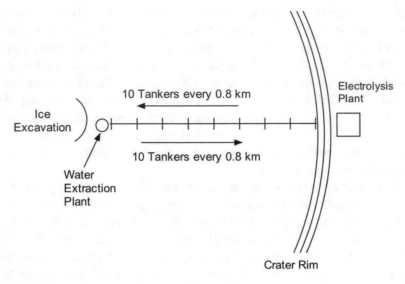

Fig. 2.2 Distribution of tankers going to and from crater rim

Table 2.8 Power levels to operate water extraction unit according to NASA

	O_2 production rate (mT/year)			
	1	10	50	100
Mass of unit	40	190	790	1520
Power (kW)	0.23	2.3	11.6	23.1

Table 2.9 Power levels to operate water extraction unit according to this book

O_2 annual production	1 mT	10 mT	50 mT	100 mT
Annual regolith rate (mT)	80	800	4000	8000
kWh	6256	62,560	312,800	625,600
Hours	3500	3500	3500	3500
kW to heat regolith	1.8	17.9	89.4	178.7

As is common in many NASA reports on ISRU, it appears that LSS10 focused on processes, while neglecting the complexity and difficulty of prospecting for resources, particularly in searching for accessible water ice in shaded polar craters. As we discussed at the end of Sect. 2.2.4, one of the main problems in utilizing polar ice on the Moon is not so much process development, but rather, prospecting for near-surface ice. What is needed is a campaign to locate, characterize and demonstrate viability of utilizing optimal deposits of near-surface ice in locations suggested by satellite observations using a neutron spectrometer. The first step in the campaign would be long-range (at least several tens of km), near-surface observations with long-range rovers carrying instrumentation such as a neutron/gamma ray spectrometer within these craters. It seems doubtful that conventional rovers would have sufficient range, and furthermore, there is a notable shortage of isotopes to power RTGs, so it is not clear how such rovers would be powered. The second step in the campaign would be to zero in on the best location found in step 1 with subsurface access to provide ground truth, determine the vertical and horizontal distribution of the ice within an optimum area, and provide information on the properties of the soil for excavation and processing. The third stage of the campaign would be to land a demonstration plant that excavates, hauls and processes regolith to obtain water, purifies the water and provides validation of this resource for ISRU. NASA does not seem to have ever proposed such a campaign. Instead, LSS10 extols the RESOLVE instrument, a conglomeration of instruments and capabilities that might not be needed until late in the campaign—if ever.

The RESOLVE instrument was described by Sanders et al. (2011c). Observations from orbit indicated the presence of water ice in crevices at the lunar poles, and RESOLVE is a payload designed to provide "ground truth" when mounted on a lander or rover. The first generation began in 2005, with second and third generation modifications made in 2007 and 2010. According to a 2014 Space News article, NASA had spent over $20 M on the project to date, and was "notionally targeted for launch in 2018, … and NASA expects its investment to top out around a quarter of a billion dollars". At this point in 2017, there doesn't seem to be any information on RESOLVE available on the Internet. If RESOLVE were mounted on a lander, it is not clear that measurements at a single point on the Moon would be representative. If mounted on a rover, the mission cost would zoom up, and the range of the rover on terrain in icy crevices might be very limited. Since lunar ISRU hardly makes sense anyway, it would seem appropriate to scrap RESOLVE and replace it with a mission for prospecting on Mars.

LSS10 also discusses the process for extracting oxygen from FeO in regolith. The process is described as "simple", operating at a mere 950 °C. They do admit however that

> … the challenges of operating a system like this in the lunar environment are significant. Not only are there significant temperature extremes, but the reactor must interact with the abrasive lunar soil repeatedly and still be able to seal against the high vacuum found on the moon.

They described two prototype reactors that were developed with NASA funding. It remains to be seen whether moving lunar solids into and out of a reactor can be done autonomously and efficiently. LSS10 went on to say:

> Unfortunately, the portion of the lunar regolith that will react with the hydrogen at these temperatures is generally below 10% and perhaps much lower at the poles where an Outpost is likely to be located, so the amount of regolith that must be processed to make a metric ton of oxygen is quite high.

In fact, it is not only "quite high", but incredibly high, as we showed in Sect. 2.3.1. Table 2.5 indicates that one must process about 1000 mT of regolith to produce 10 mT of oxygen.

The carbothermal process for extracting oxygen from silicates in regolith was also discussed by LSS10. However, as we showed in Sect. 2.3.2, this process is a nightmare in every respect.

As is usual in the NASA community, program managers spend a great deal of time making presentations and writing summaries that are quite repetitive. It seems that in order to keep a NASA program going, presentations to justify continuation must be made repeatedly. ISRU Managers spend most of their time preparing glossy PowerPoint presentations trying to convince higher level managers of the merits of ISRU. Larson and Sanders (2010) more or less duplicates LSS10. Sanders et al. (2011) provide a broader perspective on lunar and Mars ISRU. Like other such documents, it poses the issue as "make it or bring it" implying that making it is always obviously better. In this respect it neglects the costs associated with ISRU prospecting, development and implementation. Forced into a corner by high-level decisions, ISRU managers have no choice but to advocate seemingly impractical approaches such as the carbothermal process and other lunar ISRU concepts. Sanders and Larson (2011) and Larson (2011) provide summaries of lunar ISRU options.

Parenthetically, it might be worth mentioning that the accepted mode for NASA presentations regarding ISRU is that each slide in a PowerPoint presentation is filled to the brim with a great amount of text, and typically, each paragraph has a small, almost undecipherable picture attached to it around the periphery. Reading slides in a NASA presentation is tantamount to drinking water aimed at you from a fire hose.

None of these reports and publications seem to have mentioned the words: "neutron spectrometer" or "gamma-ray spectrometer"—the most important need for prospecting for hydrogen on the Moon or Mars.

During the Griffin era of extreme focus on lunar missions, the NASA ISRU community was under orders to concentrate totally on lunar ISRU. Being capable engineers, they investigated quite a number of lunar ISRU processes, and developed some of these to various levels of maturity. But the real gut questions of cost/benefit were not usually addressed, and the ruling mantra seems to have been that if we make it via ISRU, it must automatically be better than bringing it from Earth. In the case of lunar ISRU this seems not to be true.

It is difficult for an observer such as myself to gain access to NASA budgets for ISRU. My impression is that the relatively well funded years for lunar ISRU ended around 2009, and the reports by Sanders and Larson in 2010 and 2011 more or less brought a close to this era. However, in late 2017 President Trump ordered a return to the Moon. It seems that we might endure "dejas vous all over again".

Value of ISRU

3

3.1 Value of Mars ISRU

3.1.1 Reductions in IMLEO from Mars ISRU

Cutright and Ambrose (2014) wrote a generic report on economic benefits of ISRU but I could not find much in that report that seemed useful for our purposes here.

The value of ISRU is due to the reduction in the mass of propellants that must be delivered to Mars. In doing this accounting, we include:

(a) Propellants used to enter the ERV into Mars orbit.
(b) Propellants used by the ERV to depart Mars orbit for return to Earth.
(c) Propellants used to ascend from Mars to rendezvous with the ERV in orbit.

Two orbits are considered. One is a circular orbit at altitude ~ 500 km, and the other is the extreme elliptical orbit described in Sect. 1.1.4.2. The circular orbit has the advantage that the requirement for ascent propellants is less, but the requirements for propellants for orbit insertion and departure are greater. If ISRU is not utilized, the circular orbit is probably preferred. But if ISRU is utilized, the greater requirement for ascent propellants is acceptable since the propellants are made on Mars and are not delivered from Earth. The benefit of using less propellants for orbit insertion and departure are significant.

As we discussed in Section 1.1.4.2, Polsgrove et al. (2015) analyzed ascent from Mars to an elliptical orbit. The mass of the ascent capsule according to their model for a crew of four is given in Tables 1.3 and 1.4. Their estimates of propellant requirements for ascent to an elliptical orbit are given in Table 1.5.

For ascent to a circular orbit we roughly estimate that propellant requirements are about 2/3 of that for ascent to an elliptical orbit. Thus we derive the data shown in Table 3.1.

Rough estimates were made of the propellant requirements for orbit insertion and departure (in this simple approximation they are the same). Assuming that ~ 40 mT ERV

© Springer International Publishing AG 2018
D. Rapp, *Use of Extraterrestrial Resources for Human Space Missions to Moon or Mars*, Springer Praxis Books, https://doi.org/10.1007/978-3-319-72694-6_3

Table 3.1 Estimates of ascent propellant requirements

Crew size	Elliptical orbit		Circular orbit	
	Oxygen required (mT)	Methane required (mT)	Oxygen required (mT)	Methane required (mT)
4	22.7	7.0	15.1	4.7
6	33.5	11.1	22.3	7.4

Table 3.2 Reduction in IMLEO from use of ISRU (crew size = 6)

ISRU?	Orbit	Brought from Earth		IMLEO for propellants (mT)
		Ascent propellants (mT)	Orbit insertion and departure (mT)	
None	Circular	29.7	88	560
O_2 only	Elliptical	11.1	34	212
$O_2 + CH_4$	Elliptical	0	34	100

must be placed into Mars orbit, a crude estimate for propellant requirements for orbit insertion and departure is as follows:

Circular orbit: 44 mT for insertion and 44 mT for departure
Elliptical orbit: 17 mT for insertion and 17 mT for departure

The gear ratios for transfer from LEO to Mars orbit and to the Mars surface are roughly estimated to be 3 and 10, respectively. The propellant masses delivered to the Mars surface are multiplied by 10, in order to estimate IMLEO, and the proellant masses delivered to Mars orbit are multiplied by 3. Therefore we can construct Table 3.2. This table indicates that for a crew of six, an O_2-only ISRU could save about 350 mT in LEO and a complete $CH_4 + O_2$ ISRU system could save about 460 mT in LEO.

3.1.2   Oxygen-Only ISRU Versus Water-Based ISRU?

Sanders (2016) presented a comparison of oxygen-only ISRU versus water-based ISRU. The criterion he used was the LM = landed mass of ISRU system plus any propellants delivered to the Mars surface from Earth. The system with the lowest value of LM is preferred. According to Sanders' estimates, the oxygen-only system requires delivering about 1.0 mT of ISRU system plus the methane used for liftoff. His estimate for methane was 6.8 mT, bringing the total LM for oxygen-only to 7.8 mT. As we pointed out in Table 1.5, the methane requirement might be in the range 7.0–11.1 mT, depending on the crew size. For the water-based ISRU system using regolith enriched with 40% gypsum, it was estimated that LM would be about 1.6 mT, and thus according to his criterion, water-based ISRU offers about a 5-times improvement in LM versus oxygen-only ISRU. However, the criterion used here is somewhat arbitrary. If one considers the total mass that must be delivered to the Mars surface and examines the impact of ISRU on that mass, the comparison between oxygen-only ISRU versus water-based ISRU looks different. Nobody really knows what the total landed mass will be, but I will hazard a wild guess that it will be 45 mT, excluding propellants and ISRU systems. Using oxygen-only ISRU, the total mass delivered to the surface (including ISRU and propellants) is 45 + 7.8 = 52.8 mT, whereas using water based ISRU it is 46.6 mT. There is still a net benefit to water-based ISRU, but it is not so dramatic. In addition, the intent to use water-based ISRU opens up the need to do considerable prospecting for good sites; it alters priorities in site selection; and it requires considerable validation of water extraction processes on Mars. So the argument in favor of water-based iSRU is not as overwhelming as might first appear.

The report by Sanders (2016) was given in greater detail by Abbud-Madrid et al. (2016). The criteria used to justify water-based ISRU over oxygen-only ISRU are arbitrary and biased toward water-based ISRU. One criterion was the mass of propellant produced per unit mass (of ISRU system and propellants) delivered to the Mars surface. In water-based ISRU only the ISRU system is delivered to Mars, but in oxygen-only ISRU, a substantial amount of methane must also be delivered to Mars. Hence the ratio of [propellants produced/mass delivered to Mars] is about 20 for water-based ISRU and it is only about 3 for oxygen-only ISRU.

3.2   Value of Lunar ISRU

The short answer is that there appears to be no value in most applications of lunar ISRU; at least not within the framework of the plan to estanlish a lunar outpost, as pursued in the Constellation Project of 2004–5.[1]

As we have shown previously, the only lunar process that might have some value is reduction of FeO in high-FeO Mare regolith. But even here, the power requirements and mass of regolith that needs to be processed are at best marginal.

Oxygen production from silicates in regolith by the carbothermal process was described as a nightmare.

The power requirement for obtaining volatiles imbedded by the solar wind is far too high to be feasible.

Water extraction from regolith pores in permanently shadowed craters near the poles was shown to be impractical.

---

[1]https://en.wikipedia.org/wiki/Constellation_program.

3.3   Future Factors that Could Influence Mars ISRU

3.3.1   Elon Musk Cost Reduction

If the cost of delivery of materiel to the Mars surface could be substantially reduced, that would reduce the value of ISRU as compared to bringing propellants from Earth. Musk (2017) proposed "reducing the cost of delivering mass to Mars by five million percent". (He also said "We'll create a city on Mars with a million inhabitants".) These are merely two of many idiotic statements he has made in press conferences. According to this, the cost per ton would be reduced to $0.00005 \times$ (current cost). It is difficult to estimate the current cost of delivering mass to the Mars surface but a rough guess is $100 million per mT. If Musk were accurate, this cost would be reduced to $5000 per mT. At that cost, bringing propellants from Earth would be simpler, cheaper and more reliable than using ISRU. However, the details of how he would achieve this incredible cost reduction do not hold up to scrutiny. Use of reusable rockets would have hardly any effect because the rocket used to go to Mars will stay there, and the saving from reusing the ERV are minimal. Refueling on orbit is a chimera because the fuels need to be put in orbit, and there is no practial extraterrestrial source for them. We have already folded in ISRU to the current cost of a Mars mission, so Musk's proposal to add ISRU is gilding the lily. Furthermore ISRU does not reduce the cost per ton to deliver cargo to Mars. Choosing the right propellants provides no benefit because we have already chosen the right propellants. It is not clear that Musk's system architecture is much different than current architectures. In total, the only thing unique about Musk's plan is the use of a super-large launch vehicle —almost five times as big as a current "heavy lift" launch vehicle. Whether such a launch vehicle is technically, environmentally and fiscally viable, remains to be seen. All that Musk's article proves is that a prominent billionaire gets a media forum even when he talks nonsense.

3.3.2   Nuclear Thermal Propulsion

The baseline propulsion system used for trans-lunar (or trans-Mars) injection is a LOX/LH2 propulsion stage atop the launch vehicle. For lowest energy transfers from LEO toward the Moon (or Mars) using LOX/LH2 propulsion, roughly 55% (or 65%) of the mass in LEO is required for the propellant and the propulsion stage, and $\sim 45\%$ (or 35%) of the mass in LEO consists of payload that is sent on its way to the Moon (or Mars). Therefore it requires about 2.8 mass units in LEO to send 1 mass unit on its way toward Mars in the lowest energy case (longest trip time). For Mars, the actual propulsion requirements depend on several factors (e.g. the specific launch opportunity and the desired duration of the trip to Mars). One can either use a lower-energy trajectory with a trip time of typically 300–400 days that requires less propellant (appropriate for cargo transfer), or a higher energy trajectory that uses more propellant with a trip time of typically 170–200 days (appropriate for crew transfer). LOX/LH2 is the most efficient form of chemical propulsion that is available. The technology for use of LOX/LH2 propulsion for Earth departure is fairly mature.

Despite the fact that LOX/LH2 is the most efficient form of chemical propulsion, the requirement that about 2.8 mass units in LEO are required to send 1 mass unit of cargo on a slow trip toward Mars is a major factor in driving up the IMLEO for Mars missions. To partly mitigate this onerous requirement, mission planners have proposed a form of exotic propulsion to replace chemical propulsion for Earth departure in their design reference missions (DRMs). Nuclear thermal propulsion (NTP) has been conjectured as a potential alternative to chemical propulsion in Mars missions. In most cases, NTP has only been considered as an alternative for Earth departure due to the fact that large amounts of hydrogen propellant must be stored and there is concern about potential boil-off problems for longer-term use. However, the 1968 Boeing mission model indicated an intent to consider storing hydrogen for up to three years and using NTP for Mars orbit insertion and Earth return in addition to Earth departure.

NASA DRMs DRM-1 and DRM-3 were based on a presumed nuclear thermal rocket (NTR) used for Earth departure. More recently, the ESAS Report and Constellation continued to place great importance on the NTR. Use of a NTR instead of chemical propulsion for trans-Mars (or lunar) injection from LEO could significantly increase the payload fraction in LEO by doubling the specific impulse from 450 to 900 s, but the increase in payload fraction will be mitigated somewhat by the relatively high dry mass of the NTR system including a nuclear reactor and hydrogen storage.

Here, the term "dry mass" refers to the sum of the masses of the nuclear reactor and rocket assembly plus the empty hydrogen propellant tanks.

Development of the NERVA (Nuclear Engine for Rocket Vehicle Application) began in the 1950s and 1960s. NERVA was a solid-core nuclear-thermal rocket engine. Hydrogen propellant passed through and was heated by a uranium nuclear reactor, which caused the propellant to be converted to a plasma, expand rapidly, and vent out of a nozzle, producing thrust. Nuclear-thermal rockets promised greater efficiency than chemical rockets, meaning less propellant was required to do the same work as an equivalent chemical system.

In 1961–2, a preliminary design was developed and a 22.5-foot NERVA engine mockup was assembled. The first three engine ground tests were alarming failures. Funding for a flight test was delayed. NERVA was relegated to a ground-based research and technology effort in 1963. A successful ground test was carried out in 1964. A 60-minute ground test was successful in December 1967.

Use of a NTR instead of chemical propulsion for trans-Mars (or lunar) injection from LEO might significantly increase the payload fraction in LEO by doubling the specific impulse of the departure system from 450 to 850 to 900 s. However the increase in payload fraction will be limited by the increased dry mass of the NTR system including the nuclear reactor and hydrogen storage.

The Boeing 1968 study used K = 0.45. DRA-5 provided more recent estimates.[2] According to DRA-5, each cargo vehicle departing Earth orbit utilized two hydrogen tanks and one thruster. The entire propulsion system had a dry mass of 49.7 mT and the hydrogen propellant mass was 93.5 mT, indicating K = 0.53. The crew vehicle had a dry mass of 90.9 mT and the hydrogen propellant mass was 202.7 mT, indicating K = 0.45. Evidently, the mass of the core stage was less important when the hydrogen propellant load was greater. For Earth departure, NASA's DRM-1 estimated that a NTR dry mass of 28.9 mT was required for a hydrogen propellant mass of 86 mT for all three vehicles (ERV, Cargo Lander, and Crew Lander) but the Crew Lander added 3.3 mT for a shadow shield. Thus the dry mass was ∼34% of propellant mass for cargo and 37% for crew. DRM-3 (a later addendum to DRM-1) used a NTR dry mass of 23.4 mT for all three vehicles, but each of these used different amounts of hydrogen propellant (ERV used 50 mT, Cargo Lander used 45.3 mT and Crew Lander used 50 mT). The dry mass ranged from ∼47% to ∼52% of propellant mass). Robert Zubrin's Mars Direct provided a formula that would have indicated more optimistic estimates of 20 and 12 mT for the NTR dry mass of these systems (propellant masses of 86 and 45 mT), equivalent to about 23% of propellants mass. Thus, previous estimates of the propulsion dry mass ratio (dry propulsion/propellant) ranged from about 0.23–0.52. Appendix 1 provides a review of hydrogen storage systems that indicates that a hydrogen storage system would require a mass of about 30% of the mass of liquid hydrogen stored. For hydrogen masses in the range 50–90 mT, the hydrogen tank would weigh 15–27 mT. In addition the reactor and remainder of the dry propulsion system is likely to weigh 10 mT or more. Therefore, based on this (admittedly crude) evidence, it appears likely that K is expected to range from about 0.4 for 90 mT of hydrogen to about 0.5 for 50 mT of hydrogen. However, since there appears to be considerable uncertainty regarding the NTR dry mass ratio, K was treated parametrically by Rapp (2015).

In addition, it is not clear how the large amount of hydrogen propellant would be stored and maintained. The volumes of liquid hydrogen implied by these large masses (86 and 45.3 mT) are 123 and 65 cubic meters.

Since there is considerable uncertainty regarding the NTR dry mass percentage, we can treat this in terms of a parameter:

$$(dry\,mass) = K\,(propellant\,mass)$$

where K is an unknown parameter that might possibly be somewhere between 0.2 and 0.6.

In addition to the problem of the NTR dry mass, another problem in the use of the NTR is the fact that for safety reasons, both real and imagined, it seems likely that public policy will require that the NTR be lifted to a higher Earth orbit before it is turned on. However, this requires that the launch vehicle burn more propellants, and therefore the net benefit of

---

[2]Bret Drake, editor (2009) "Human Exploration of Mars—Design Reference Architecture 5.0" NASA Report SP 2009-566.

Table 3.3 Estimated reduction in payload mass lifted to circular Earth orbit versus altitude

Altitude (km)	% Reduction from 200 km
200	0
250	1.9
490	9.8
750	16.5
1000	21.1
1250	24.0

using the NTR will be reduced compared to firing it up in LEO. In fact the ESAS Report indicated that it would be lifted to 800–1200 km altitude rather than the typical starting point of ~200 km altitude of LEO. The estimated reduction in payload lifted to various Earth orbit altitudes is given in Table 3.3. For example, the payload mass lifted to 1000 km altitude is about 80% of that which could be lifted to 200 km altitude.

Table 3.4 shows the fraction of mass originally Earth orbit that can be sent on a fast trajectory to Mars in 2022. Using chemical propulsion and departing from 200 km LEO, 31% of the mass in LEO can be sent toward Mars. The fraction of mass originally in a 200 km LEO that can be sent on a fast trajectory to Mars in 2022 using NTR propulsion depends on K and the altitude of start-up. For K ~ 0.5 and start-up at 1250 km, the NTR would produce a very small improvement (33%). This small improvement would come at an enormous cost to develop, test and validate the NTR. On the other hand if K ~ 0.3 and start-up occurs at 1000 km, 42% of the mass originally in a 200 km LEO can be sent on a fast trajectory to Mars in 2022. This represents a 35% increase in payload sent toward Mars using NTR. The benefit/cost ratio is uncertain until we can pin down these parameters more precisely. Nevertheless, NASA has resolutely included the NTR in its *Design Reference Missions*.

In a NASA presentation in late 2006, it was proposed to utilize the NTR not only for Earth departure, but also for Mars orbit insertion and Earth return from Mars orbit. This would entail storing multi-tens of mT of hydrogen for two to three years in space. Exactly how this would be accomplished was not revealed.

In most cases, NTP has only been considered as an alternative for Earth departure due to the fact that large amounts of hydrogen propellant must be stored and there is concern about potential boil-off problems for longer-term use. However, Anonymous (2006) has

Table 3.4 Ratio of initial total mass in Earth orbit to payload sent toward Mars

Altitude at start-up (km)	Chemical propulsion	NTR with various K				
		0.2	0.3	0.4	0.5	0.6
200	3.23	1.82	1.96	2.13	2.33	2.50
1000		2.22	2.38	2.56	2.78	3.03
1250		2.38	2.56	2.78	3.03	3.33

These figures pertain to a fast flight at the 2022 opportunity indicated an intent to consider storing hydrogen for up to three years and using NTP for Mars orbit insertion and Earth return in addition to Earth departure.

3.3.3 Solar Electric Propulsion

3.3.3.1 Orbit Raising

One of the major requirements for propellants in any space exploration venture is to depart from the gravitational influence of the Earth. While most mission concepts call for an Earth departure propulsion stage to fire up in LEO, another conceivable alternative is to use solar electric propulsion (SEP) to raise the orbit of the spacecraft from LEO to a high Earth orbit, and thereby reduce the propulsion requirements for Earth departure from this high orbit. The question then arises as to the requirements for the solar electric propulsion system, and whether such a system is feasible and cost-effective. To the extent that orbit raising with SEP reduces the cost of transfer of mass from Earth to Mars that would reduce the value of ISRU.

In a series of *Design Reference Mission* studies for human missions to Mars, NASA had to deal with the inevitable problem of sending large masses to Mars, which required much larger masses in LEO. During the late 1990s, their studies utilized a NTR for Earth departure from LEO to reduce the IMLEO. However, in the "Dual Landers" study ($\sim$2000) they eliminated the NTR and used SEP for orbit raising instead. It is not clear why this change was made but presumably it was because of uncertainties in the political viability of firing up a NTR in LEO, as well as concerns about the cost of developing the NTR.

In the Dual Landers concept, a "tug" powered by solar electric propulsion is used to lift space vehicles from LEO to an elongated elliptical Earth orbit because trans-Mars injection with chemical propulsion requires far less propellant from this orbit than it would from LEO. The energy that would have been used for departure from LEO using chemical propulsion is mostly replaced by solar energy that drives the electric propulsion system used to raise spacecraft to a high orbit, although the mass of Xenon propellant is about 80% of the payload delivered to high orbit. The documentation of the Dual Landers mission is sparse[3] and it is difficult to appraise the feasibility of the masses used in this mission concept. The slow spiraling out of the SEP tug (several months required for transfer) creates time delays and operational scheduling difficulties. Because the SEP tug drags the vehicles slowly through the radiation belts, the crew would have to wait until the *Trans-Habitat Vehicle* reached HEO before using yet another vehicle, a fast "crew taxi" to rendezvous with the Trans-Habitat Vehicle.

---

[3]As far as I know, the "Dual-Landers" study is not available on the Internet, although a one-page bulletized synopsis is available at http://history.nasa.gov/DPT/Architectures/Recent%20Human%20Exploration%20Studies%20DPT%20JSC%20Jan_00.pdf. I was once lucky to borrow a dog-eared hard copy for a few days.

Woodcock (2004) described a hypothetical SEP system for orbit raising heavy loads. This reference utilized a payload of 50 mT driven by a 500 kW solar electric propulsion system with a specific impulse of 2000s. The trip time (up) was 240 days and (down) was 60 days. The critical parameters of the propulsion system were estimated as:

Thruster mass = 2 kg/kW
Power processing mass = 4 kg/kW
Array mass = 143 W/kg = 250 W/m^2
Array areal density = 1.8 kg/m^2.

These are ambitious figures but may possibly be achievable some day. The masses of various elements are summarized in Table 3.5. The required amount of Xe propellant per transfer is 41.2 mT.

The lighting industry utilized 3×10^6 L of Xe during 2013. This added to the previous usage of about 10×10^6 L/year. Xenon prices fluctuated rather wildly over the past decade, reaching a high of \$25/l in 2008, and falling as low as \$5/l in 2011. Projections for the future[4] suggest demand will outstrip the supply by about 20% over the next few years. Prices are likely to be in the \$20/l range.

With current world production of 13×10^6 L/year = 69 mT/year, one transfer would require approximately 45% of the present annual world production of Xe. Furthermore, if Xe costs about \$20/l, the cost of Xe for one orbit transfer could be \$155 M. While it might be possible to increase world production significantly, recent articles suggest difficulties. The viability of the SEP tug concept depends critically on use of a hypothetical high-efficiency lightweight solar array, and lightweight propulsion components. Furthermore, it seems very uncertain whether the required amount of Xenon propellant could be obtained, and if obtainable, whether the cost would be affordable. Radiation would gradually diminish the efficiency of the solar arrays with each passage through the radiation belts. The total cost of the system includes the SEP tug, the mission operations involved, and the fast transit vehicle to take the crew up to high orbit for rendezvous with the *Trans-Habitat Vehicle*. Because of the long time required for transfer, several of these "tugs" might be needed.

At this point, it seems unlikely that such a scheme would be viable.

The data in Table 3.5 show that the gear ratio for delivery of mass from LEO to high orbit (assume GEO) is 110,451/50,000 = 2.21. Since the required Δv for escape from GEO is 1.27 km/s, the gear ratio for escape from GEO is roughly 1.4 (Curtis 2005). Hence for the first operation of the SEP tug, the gear ratio for sending a payload toward Mars is $2.21 \times 1.4 = 3.1$. This is similar to that using propulsion from LEO. Since the SEP system is reusable, the mass that must be added in LEO in each subsequent operation drops somewhat, but not nearly enough to make this approach attractive.

---

[4]Richard Betzendahl (2014) "The Rare Gases Market Report" http://www.cryogas.com/pdf/Link_2014RareGasesMktReport_Betzendahl.pdf.

Table 3.5 Estimated masses of SEP orbit-raising system for 50,000 kg payload

Up trip time	240 d
Return trip time	61 d
Array area	2000 m^2
Payload accommodation mass	5000 kg
Array mass	2500 kg
Thruster mass	1000 kg
PPU & cabling mass	2000 kg
Propellant tank mass	2060 kg
Structures mass	4483 kg
Inert mass	19,255 kg
Return cutoff mass	19,255 kg
Up propellant	31,394 kg
Return propellant	7841 kg
Unusable propellant	1962 kg
Total propellant	41,196 kg
Total initial mass in LEO	110,451 kg

Bonin and Kaya (2007) carried out an extensive study of the merits of orbit-raising for interplanetary missions using SEP. They pointed out that because SEP requires considerable time, cryogenic propellant boil-off is a detriment to orbit-raising with SEP and must be taken into account. Time is also a detriment in itself. They calculated the Δv for escape from elongated elliptical axes as a function of semi-major axis length and obtained the results shown in Fig. 3.1. Beyond 8500 km, the rate of fall-off of Δv with semi-major axis slows down. Most of the reduction in Δv is achieved by 20,000 km. There is little need to raise the orbit beyond 20,000 km.

Fig. 3.1 Earth departure Δv toward Mars as a function of the Earth orbit semi-major axis

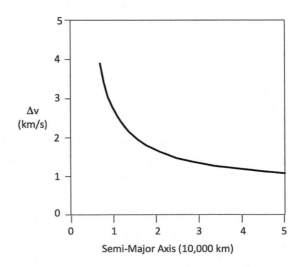

The analysis provided by Bonin and Kaya is very intricate and has many nuances. Surprisingly, they did not refer to Woodcock's work, although their estimates for specific power and densities of the arrays were similar to those of Woodcock. It is difficult to summarize all their work. Only a few conclusions are reported here. They found that propellant boil-off, the reserve fuel required for tug return to LEO, and power degradation can severely limit the relative payload gain of SEP. Economic considerations of propellant procurement might also be a show stopper. Although they recommended further study, the case for SEP orbit-raising seemed very marginal.

3.3.3.2 SEP Cargo Transfer

Over the past few years, several investigators have modeled transfer of cargo to Mars via solar-electric propulsion (SEP). The Aerojet Corporation has been a leader in this endeavor (e.g. Cassady et al. 2015; Carpenter et al. 2012). While the virtues of using SEP have been emphasized, there has been a notable lack of detail provided regarding the components of the system and the trajectories. Data parallel to Table 3.5 does not seem to be available in the various publications. It appears that use of SEP could reduce IMLEO for cargo delivery to Mars as compared to chemical propulsion, but at this point it is unclear what the quantitative benefit would be. On the other hand, use of SEP would require use of a large amount of Xenon, and the trip time would longer than the 26-month Mars delivery cycle, which would introduce logistic problems for a mission. From what can be pieced together from the fragmentary reports in the literature, the mass saving from use of SEP would not be great enough to affect the value of ISRU very much.

3.3.4 Use of Aero-Assist for Mars Orbit Insertion and Descent

An all-propulsive approach for entry, descent and landing (EDL) at Mars leads to very high entry system masses (Marsh and Braun 2009).

Typical Mars design reference missions (DRMs) assumed use of aerocapture to insert assets into Mars circular orbit, aero-assisted descent and precision landing to place assets on the Mars surface. (NASA DRM-1, NASA DRM-3, ESAS, Constellation) An *Earth Return Vehicle* (ERV) is placed into Mars orbit while landed assets descend to the surface. On the return leg of the mission, the *Mars Ascent Vehicle* (MAV) performs a rendezvous with the ERV, transferring the crew to the ERV for Earth return. Some DRMs (Mars Direct, Mars Society) were based on direct entry for landed assets without passing through the orbit insertion step, but based on recent NASA planning documents, it seems likely that NASA will require preliminary orbit insertion prior to descent in order to reduce perceived risk and add mission flexibility. (ESAS Constellation) The Mars Direct DRM utilized direct return from the Mars surface to Earth without a rendezvous in Mars orbit, but this inherently requires a very large capacity in situ resource utilization (ISRU) plant built into the campaign to produce typically ~ 100 mT of propellants, and NASA plans for utilizing ISRU appear to be less ambitious. In addition, in the current NASA ideology,

ISRU might be added late in the campaign to a system based on non-use of ISRU, and would therefore not change the masses of vehicles (ESAS, Constellation).

In order to implement a human mission to Mars, it is assumed here that some assets (mass = M_{MO}) are placed into Mars orbit, and some assets (mass = M_{MS}) are transported to the surface after first being placed into Mars circular orbit and descending from there.

There is no experimental database on which to base estimates of requirements for entry and descent systems for aerocapture by very large payloads. Therefore, DRMs have had to rely upon projections based on models not supported by experimental data. All of the NASA DRMs made very optimistic assumptions regarding the required mass for entry, descent and landing (EDL) systems for Mars prior to the important analyses by Braun and the Georgia Tech team and Manning of JPL around 2006.

It is instructive to review the original aerocapture design for the MSP 2001 Orbiter, even though it was much smaller than systems for human missions, and therefore the results are not necessarily directly transferable to human mission scale. Various parameters associated with this system are listed in Table 3.6. The data in Table 3.6 were taken from a MSP 2001 orbiter report that no longer seems to be available on the Internet.

It can be seen that the critical ratio:

$$(\text{Entry system mass})/(\text{Mass placed into Mars orbit}) = 0.6,$$

and this implies that the gear ratio is:

$$G = (\text{Approach mass})/(\text{Mass placed into orbit}) = 1.6.$$

If the cruise stage had been included in the approach mass, this latter ratio would have increased to 1.98.

As Braun and Manning (2006) pointed out, the United States had successfully landed five robotic systems on the surface of Mars by that time. These systems all had landed mass less than 600 kg (more recently the NASA Mars Science Laboratory (MSL) upped that to about 900 kg). The MSL Mass was 3893 kilograms total at launch, consisting of a 899-kg rover, a 2401-kg entry, descent and landing system (aeroshell plus fueled descent stage), and a 539-kg fueled cruise stage.

For human missions to Mars, vehicles of mass > 40 mT will likely have to be landed.

According to NASA mission concepts, payloads will be inserted into Mars circular orbit by aerocapture prior to descent to the surface. Since several payloads must be landed at the same site, pinpoint landing to within 10–100 m is also required. Compared to past robotic landers, human missions require a simultaneous two order of magnitude increase in landed mass capability, four order of magnitude increase in landed accuracy, and entry, descent and landing (EDL) operations that may need to be completed in a lower density (higher surface elevation) environment.

The major challenge in developing aero-assist systems for Mars is the thinness and variability of the atmosphere. In order to achieve rapid deceleration at sufficiently high

Table 3.6 Some characteristics of the MSP 2001 orbiter aerocapture system (as of the end of Phase B of the mission)

Parameter	Value
Maximum deceleration	~4.4 g's at 48 km altitude
Minimum altitude	~48 km
Time duration at altitude < 125 km	~400 s
Mass prior to Mars entry	544 kg
Heatshield mass	122 kg
Backshell mass	75 kg
Total entry system mass (includes all structures and mechanisms)	197 kg
Entry system mass/mass prior to Mars entry	0.36
Entry system mass/mass placed into orbit	0.60
Approach mass/mass placed into orbit	1.6
Post-Aerocapture periapsis-raise maneuver propellants (400 km circular orbit)	20 kg
Payload mass in Mars orbit	327 kg

altitudes that enough altitude remains for final descent with parachutes, the aerodynamic forces must be as large as possible compared to the inertial force of the mass being decelerated. Braun and the Georgia Tech team carried out simulation models of the orbit insertion and descent processes for very large vehicles (Wells et al. 2006). In a more recent study, the Georgia Tech team updated their preliminary results (Christian et al. 2006).

After aerocapture into orbit, a vehicle delivering humans or cargo to the surface must perform entry, descent and landing. Christian et al. (2006) considered two options: (1) a single heat shield used for both aerocapture orbit insertion and descent, and (2) a dual heat shield set in which one shield is used for aerocapture orbit insertion and jettisoned prior to descent using the second shield.

The single heat shield approach is appealing in its simplicity. However, Braun and Manning (2006) raised three concerns about this approach. First, since the *Thermal Protection System* (TPS) must be sized for the harsher aerocapture environment, the vehicle performs its entry from orbit with a more massive, higher ballistic coefficient heat shield than would nominally be required for descent. Second, if the vehicle does not jettison the aerocapture heat shield following the aerocapture maneuver, it must be designed to withstand a large amount of heat soaking back into the vehicle structure from the thermal protection system. Finally, a third challenge results from the fact that the vehicle, having been inserted into Mars orbit, must operate functionally prior to descent. It is not clear how to accommodate power, thermal, orbit-trim propulsion, communications, and other spacecraft functions needs from within the confines of the aeroshell.

The alternate approach for the TPS configuration would be to use separate, nested heat shields for aerocapture and EDL. This provides the benefit of jettisoning the hot aerocapture TPS immediately following the aerocapture maneuver, and allows the use of much lighter TPS for entry, thus minimizing the vehicle's ballistic coefficient for that maneuver. The disadvantage of this approach is that packaging two nested heat shields on the vehicle requires a means of securing the primary heat shield to the structure and separating it without damaging the secondary heat shield and would likely result in a significant overall mass penalty.

Analyses of entry/descent systems were made by the Georgia Tech team for 10 and 15 m diameter heat shields, using single or dual heat shields, and using propulsive descent with or without assistance from a 30 m parachute. (Wells et al. 2006; Christian et al. 2006) The mass of the entire entry, descent and landing (EDL) system was computed by combining the estimated masses of each of the major EDL subsystems (heat shield, LOX/CH4 main propulsion system, reaction control system (RCS), backshell and parachute). Additionally, an EDL mass margin of 30% of the entry system dry mass (17% of total entry system wet mass) was included.

The figure of merit for the study was:

$$\text{payload fraction} = (\text{landed payload mass})/(\text{initial approach mass})$$

This is the reciprocal of the gear ratio for transfer from approaching Mars to landing on the surface.

The Georgia Tech team compared results for L/D = 0.3 and L/D = 0.5 entry vehicles, and they found that the lower L/D can land slightly larger payloads on the Martian surface. In general, it was found that use of a single heat shield resulted in better figures of merit compared to use of dual heat shields.

An important question is whether it is feasible to use parachutes in the descent sequence for human-scale Mars missions. The study concluded: "material strength and stability concerns led to a limit on the parachute diameter of 30 m." All of the cases investigated in the study pushed up against this limit and required a 30 m diameter parachute. The payload fraction was typically slightly higher using a parachute.

The results indicated that the payload fraction is higher (up to 0.30) using the smaller (10 m) entry vehicle, as compared to a payload fraction of about 0.25 using the larger (15 m) entry vehicle. However, it was impractical to use the smaller entry vehicle for an initial approach mass greater than 60–80 mT (landed payload mass $\sim$18–22 mT). Using the larger entry vehicle (15 m diameter), an initial mass of $\sim$100 mT can land 25 mT of payload. By contrast, in some previous DRMs, the payload fraction was taken as $\sim$0.7, leading to much lower estimates of IMLEO (DRM-1, DRM-3). Unfortunately, this assumption of a higher payload fraction was based on analyses that were not published or released to the public, so there is no way to compare the NASA models to the Georgia Tech models. It seems probable that such NASA models were sketchy and not done in the depth of the Georgia Tech studies. One contribution to this difference is that Georgia Tech tacked on a 30% mass margin to the entry system dry mass, and NASA might not have included such a factor.

If the Georgia Tech model for the 15 m shield is used, the gear ratio G $\sim$4.0. However, this included an allowance of 30% of dry mass (17% of wet mass) for entry system mass margin. If we arbitrarily cut this margin to 11% of wet mass, the value of G $\sim$3.6 and we will use this approximation. It should be emphasized that the MSP'01 data were not used in deriving this result. It derives entirely from the Georgia Tech model.

The conclusion that can be drawn from this work is that there is likely to be an upper limit to the mass of any one unit that can be delivered to the Mars surface, and this is likely to be in the range of $\sim$25 mT and the gear ratio for delivery of this mass is estimated to be roughly G $\sim$3.6. Steinfeldt et al. (2009) studied the potential benefits of inflatable aerodynamic decelerators, vehicle shape, and supersonic retropropulsion in allowing more massive payloads to be landed. They concluded that a 20 mT payload could be landed using a 10 m aeroshell with an entry mass $\geq$ 40 mT. A 37 mT payload would require an entry mass > 100 mT.

The Georgia Tech study did not divide the EDL system into separate parts for aerocapture into orbit and subsequent descent to the surface. It seems likely that the majority of the heat shield and aeroshell masses are attributable to orbit insertion whereas the majority of propulsion mass (and of course the parachute) would be attributable to descent. Based on the detailed mass breakdown provided by Georgia Tech it has been roughly estimated that the total EDL system mass is roughly 52% of approach mass for aerocapture orbit insertion, and 48% of mass in orbit for descent and landing from orbit. Since the total EDL system mass for orbit insertion and descent and landing was estimated to be roughly 72% of a large mass approaching Mars, the mass requirements for entry systems are estimated as follows. For a 100 mT mass approaching Mars, about 28 mT can be landed on Mars. The remaining 72 mT consists of two subordinate entry systems, with $\sim$37 mT allocated to the aerocapture orbit insertion system and $\sim$35 mT allocated to a descent and landing system. Thus, a 40 mT entry system can place 34.6 mT into Mars orbit, but 19 mT of the mass in orbit represents the descent and landing system, and the payload delivered to the Mars surface is 15.6 mT. It is concluded that for large masses, the entry system mass for orbit insertion is estimated to be roughly 21/34.6 $\sim$ 60% of the mass inserted into orbit, and the entry system for orbit insertion and descent and landing is about 40/15.6 = 256% of the landed payload mass. The estimated value for the gear ratio for (Trans-Mars $\rightarrow$ Mars orbit) using aerocapture is 1.6.

As Rapp (2015) showed, the expected value for the gear ratio G (Trans-Mars $\rightarrow$ Mars surface) using LOX-CH4 propulsion (at this point a likely possibility) is around 9.3. According to Georgia Tech results, the expected value of G using aero-assist is 3.6. Thus, use of aero-assist reduces G (and therefore IMLEO) by a factor of 9.3/3.6 $\sim$ 2.6 compared to use of chemical propulsion for MOI. However, if the payload fraction was as high as 0.7 (as was assumed by NASA DRMs) the value of G would be $\sim$1.4 and use of aero-assist would reduce IMLEO by a factor of $\sim$7 compared to chemical propulsion.

Another important consideration is how much volume is available within the aeroshell. The Georgia Tech study concluded that pressurized volumes will be $< \sim 60$ and $287 \ m^3$ in the 10 and 15 m diameter entry vehicles, respectively. A significant portion of this pressurized volume will be taken up by cargo, crew accommodations, and other related subsystems, resulting in a habitable volume that is noticeably smaller. The study concluded that while such a small capsule might be allowable for the short descent and ascent transits, it would be very inadequate as a long-term habitat on the surface. They concluded that a dedicated habitat is needed for a human Mars mission. However, a rigid habitat of sufficient size cannot fit within the capsule shape. An inflatable habitat would likely be necessary. If ISRU is used for oxygen production only, and liquid methane is brought from Earth, the volume problem would be increased significantly.

Use of aero-assist for orbit insertion and descent and landing will reduce the required propellant mass in LEO compared to an all-propulsive approach. That, in turn, will reduce the value of ISRU. Alternatively, if Elon Musk can significantly reduce the cost of an all-propulsive approach, that would also reduce the value of ISRU.

Adler et al. (2009) drafted a well thought out, fully rational plan for developing large-scale EDL systems for Mars. It seems to have been widely ignored by the NASA management. As is usual in NASA, a necessary, albeit expensive technology program needed to fulfill the hype in NASA PowerPoint presentations of the "Evolvable Mars Campaign" is unfunded while NASA continues to make bold claims ("Mars in the 2030s") without developing the needed technology.

Refueling Spacecraft in LEO Using Propellants Derived from the Moon or Asteroids

4

4.1 Introduction

If Mars-bound vehicles could be fueled in LEO with H_2 and O_2 from the Moon, the required mass of Mars-bound vehicles to be delivered from Earth to LEO would be reduced to about 40% of the required mass if propellants were brought up from Earth. For example, a Mars-bound vehicle that weighs say, 250 metric tons in LEO when propellants are brought from Earth, if fueled by hydrogen and oxygen from the Moon would have a mass of only about 100 metric tons. This would have a huge impact on the feasibility of launching large Mars-bound vehicles.

Using an extension of a model developed previously, the percentage of water mined on the Moon that can be transferred from the Moon to LEO for fueling Mars-bound vehicles can be roughly estimated. It is assumed here implicitly that accessible water can be exploited on the Moon. If that is not the case, this entire concept becomes moot. Furthermore, the process may become untenable if the vehicle masses are too high.

We assume that a system is in place to extract water on the surface of the Moon. Part of the water mined on the Moon is electrolyzed to produce H_2 and O_2 propellants for transporting water to Lunar Lagrange point #1 (LL1). LL1 is an interesting place because it takes very little propellant to get there from lunar orbit or high Earth orbit (see "L1" in Fig. 4.1). It has sometimes been suggested that propellant depots should be placed at LL1 and staging of interplanetary spacecraft carried out at this junction.

Two vehicles are used in this process. A Lunar Water Tanker (LWT) carries water from the lunar surface to LL1. At LL1, part of this water is electrolyzed to provide propellants to return the LWT to the Moon, and part is electrolyzed to provide propellants to send the LL1-to-LEO Tanker (LLT) to LEO. The remaining water that is not electrolyzed is carried to LEO. At LEO, this water is electrolyzed and part of the H_2 and O_2 is used to return the LLT to LL1. The remainder is used as propellants to fuel up a Mars-bound vehicle in LEO

© Springer International Publishing AG 2018
D. Rapp, *Use of Extraterrestrial Resources for Human Space Missions to Moon or Mars*, Springer Praxis Books, https://doi.org/10.1007/978-3-319-72694-6_4

Fig. 4.1 Earth-Moon lagrange points (not to scale). LEO and GEO are low Earth orbit and geostationary Earth orbit

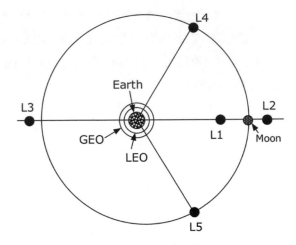

Fig. 4.2 Outline of process for transporting water from Moon to LEO

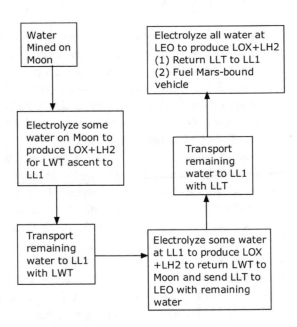

(see Fig. 4.2). A figure of merit is the net percentage of water mined on the Moon that can be transported to LEO for use by Mars-bound vehicles.

The Δv values for various orbit changes provided by Blair et al. (2002) are listed in Table 4.1. The value for LL1-LEO requires aerocapture at LEO.

Table 4.1 Estimated Δv (m/s) for various orbit changes	Transfer	Δv (m/s)
	Earth–LEO	9500
	LEO–GEO	3800
	GEO–LEO with aerobraking	500
	GEO–LL1 (assumption only)	800
	LL1–LEO with aerocapture	500
	LEO–LL1	3150
	LL1–LLO	900
	LL1–lunar surface	2390
	Lunar surface–LL1	2500

4.2   Value of Lunar Water in LEO

A major impediment to viable scenarios for human exploration of Mars is the need for heavy vehicles that must be landed on Mars. For each kg of payload landed on Mars, it may require about 9–11 kg in LEO, depending on systems used for entry, descent and landing, assuming that hydrogen/oxygen propulsion is used for trans-Mars injection (TMI) and full aero-assist is used at Mars. Thus, a 40 metric ton Mars lander would require perhaps 360 (or more) metric tons in LEO.

About 60% of the mass in LEO consists of H_2 and O_2 propellants for trans-Mars injection out of LEO. If such Mars-bound vehicles could be fueled in LEO with H_2 and O_2 from the Moon, the required mass of Mars-bound vehicles to be delivered from Earth to LEO would be reduced to about 40% of the mass that would be required if propellants were brought from Earth.

4.3   Percentage of Water Mined on the Moon Transferred to LEO

The percentage of the water mined on the Moon that can be transferred to LEO depends critically on the masses of the vehicles used for transport. While Blair et al. (2002) were concerned with a different issue: commercialization of orbit-raising communication satellites, the mass and propellant analyses are directly transferable to our concern: fueling Mars-bound vehicles in LEO with lunar-derived propellants.

4.3.1   Transfer via LL1

Blair et al. made optimistic assumptions for the masses of transfer vehicles. In the present analysis, the computations of Blair et al. are generalized by allowing the masses of the

LWT and the LLT to vary widely as parameters. Using the Δv estimates of Table 4.1, the efficiency of transfer of water mined on the Moon to LEO is estimated as a function of the tanker masses.

Electrolysis of water produces 8 kg of O_2 for every kg of H_2. Since the optimum mixture ratio for O_2/H_2 propulsion is assumed to be 6.5:1, 1.5 kg of excess O_2 will be produced per kg of H_2 that is produced. This O_2 would likely be vented, although O_2 in LEO might be useful for various purposes such as human life support. This indicates that per kg of water electrolyzed, only $7.5/9 = 0.833$ kg of useful propellants are produced.

The mass of either vehicle (LLT or LWT) is represented as a sum of three masses:

M_p = propellant mass
M_i = inert mass (including the structure, an aeroshell for the vehicle that goes to LEO, a landing system for the vehicle that goes to the lunar surface, the water tank, the (reusable) propulsion stage, and the avionics).
M_w = water mass carried by the vehicle to the next destination
M_t = total mass = $M_p + M_i + M_w$

The inert masses of the LWT and LLT tankers are of critical importance in this scheme. We shall assume that the inert mass is some fraction of the total mass:

For each vehicle, we set

$$M_i = K\,(M_t)$$

where K is an adjustable parameter, and we define K_1 for the LWT and K_2 for the LLT independently.

We begin the calculation by assuming that we will extract enough water on the Moon to send 25 mT of water to LL1. We will then work backwards to estimate how much water would have to be extracted on the Moon in order to provide propellants to send 25 mT of water to LL1. These results can be scaled to any arbitrary amount of water transferred to LL1.

The rocket equation provides that

$$M_p/(M_i + M_w) = R - 1$$
$$M_t/(M_w + M_i) = (M_p + M_w + M_i)/(M_w + M_i) = R$$
$$M_t/M_p = R/(R - 1)$$

For transfer from the lunar surface to LL1, we have

$$R = \exp\left(\Delta v/(9.8 \times I_{sp})\right) = \exp\left(2500/(9.8 \times 450)\right) = 1.763$$

The total mass on the lunar surface is

$$M_t = M_p + M_i + M_w$$
$$M_t = M_t(R - 1)/R + K_1 M_t + M_w$$
$$M_w = M_t[1 - (R - 1)/R - K_1]$$
$$M_t = M_w/[1 - (R - 1)/R - K_1]$$

Since we have specified $M_w = 25$ mT, we can calculate M_t. From this, all the other quantities can immediately be calculated. Table 4.2 shows the calculations for the transfer from the Moon to LL1.

The calculation process occurs as follows for any row with stipulated values of R in col. J and K_1 in col. L.

Step 1: Put 25.0 into col. H as the water transferred to LL1.
Step 2: Calculate the total mass in col. C.
Step 3: Calculate inert mass in col. D.
Step 4: Calculate propellant mass in col. E.
Step 5: Calculate the mass of water that is electrolyzed in col. F.
Step 6: Calculate the mass of excess oxygen produced in col. G.
Step 7: Calculate the mass of water mined on the Moon in col. I.

The next step is returning the empty LLT from LL1 to the Moon. The spreadsheet for doing this is shown in Table 4.3. One must use the same value of K_1 in Table 4.3 as was used in Table 4.2. This dictates the row to be used in Table 4.3. Negative values in Column H for water remaining at LL1 indicate that for sufficiently high values of K_1, no water can be transferred from the Moon to LL1.

The procedure is as follows:

Step 1: Use the values of inert mass from col. D in Table 4.2 in col. D of Table 4.3. The inert mass is the same in both tables.
Step 2: Calculate the propellant mass in col. E.
Step 3: Calculate the total mass of the returning LLT as the sum of the inert mass plus the propellant mass in col. C.
Step 4: Calculate in col. F the amount of water that must be electrolyzed at LL1 to produce the required propellant in col. E.
Step 5: Calculate excess oxygen produced by electrolysis in col. G.
Step 6: Calculate the amount of water remaining at LL1 after departure of the LLT in col. H.

The next step is transfer of water from LL1 to LEO. Here, a trial-and-error procedure is used. We guess how much water can be transferred to LEO, and the propellant requirements for transfer from LL1 are calculated for this load, assuming some value of K_2. The amount of water that must be electrolyzed at LL1 to produce propellants for this transfer to

Table 4.2 Sample spreadsheet for calculating requirements to transfer 25 mT of water from the Moon to LL1 as a function of K_1

Row	C	D	E	F	G	H	I	J	K	L
1	Columns									
2	Total mass	Inert mass	Propellant mass	Water electrolyzed	Excess O_2	Water transferred	Water mined on Moon	Rocket equation factors		
3	M_t	M_i	M_p	M_{el}	M_{xs}	M_w	M_M (mined)	R	R − 1	K_1
4	= H/[1 − L − (K/J)]	= L * C	= C − D − H	= 1.2 * E	= F − E		= H + F		= J − 1	
5	53.50	5.35	23.15	27.78	4.63	25.00	52.78	1.763	0.763	0.10
6	58.51	8.19	25.32	30.38	5.06	25.00	55.38	1.763	0.763	0.14
7	64.55	11.62	27.93	33.52	5.59	25.00	58.52	1.763	0.763	0.18
8	71.99	15.84	31.15	37.38	6.23	25.00	62.38	1.763	0.763	0.22
9	81.36	21.15	35.20	42.25	7.04	25.00	67.25	1.763	0.763	0.26
10	93.53	28.06	40.47	48.57	8.09	25.00	73.57	1.763	0.763	0.30
11	109.99	37.40	47.60	57.12	9.52	25.00	82.12	1.763	0.763	0.34
12	133.49	50.72	57.76	69.31	11.55	25.00	94.31	1.763	0.763	0.38
13	169.74	71.29	73.45	88.14	14.69	25.00	113.14	1.763	0.763	0.42

Table 4.3 Sample spreadsheet for calculating requirements to return the LLT from LL1 to the Moon

Row	C	D	E	F	G	H	J	K	L
Columns	Total mass	Inert mass	Propellant mass	Water electrolyzed	Excess O_2	Water remaining	Rocket equation factors		
	M_t	M_i	M_p	M_{el}	M_{xs}	M_w	R	R − 1	K_1
	= D + E	= D(T₂)	= D * K	= 1.2 * E	= F − E	= 25 − F		= J − 1	
5	9.20	5.35	3.85	4.62	0.77	20.38	1.719	0.719	0.10
6	14.08	8.19	5.89	7.07	1.18	17.93	1.719	0.719	0.14
7	19.98	11.62	8.36	10.03	1.67	14.97	1.719	0.719	0.18
8	27.23	15.84	11.39	13.67	2.28	11.33	1.719	0.719	0.22
9	36.37	21.15	15.22	18.26	3.04	6.74	1.719	0.719	0.26
10	48.25	28.06	20.19	24.22	4.04	0.78	1.719	0.719	0.30
11	64.30	37.40	26.90	32.28	5.38	−7.28	1.719	0.719	0.34
12	87.21	50.72	36.49	43.79	7.30	−18.79	1.719	0.719	0.38
13	122.57	71.29	51.28	61.54	10.26	−36.54	1.719	0.719	0.42

"D(T2)" means D from Table 4.2

Table 4.4 Sample spreadsheet for calculating requirements to transfer water from LL1 to LEO

Row	Columns										
	C	D	E	F	G	H	I	J	K	L
	Total mass	Inert mass	Propellant mass	Water electrolyzed	Excess O_2	Water transferred	Water transferred	Rocket equation factors		
	M_t	M_i	M_p	M_{el}	M_{xs}	M_w (guess)	M_w (calc.)	R	R − 1	K_2
	=H/[1 − L − (K/J)]	= L * C	= C − D − H	=1.2 * E	= F − E	$H(T_3)$	= H − F		= J − 1	
5	22.12	2.21	2.37	2.85	0.47	17.54	17.54	1.12	0.12	0.1
6	19.46	1.95	2.09	2.50	0.42	15.43	15.43	1.12	0.12	0.1
7	15.26	1.53	1.64	1.96	0.33	12.10	13.01	1.12	0.12	0.1
8	12.29	1.23	1.32	1.58	0.26	9.74	9.75	1.12	0.12	0.1
9	7.30	0.73	0.78	0.94	0.16	5.79	5.80	1.12	0.12	0.1
10	0.83	0.08	0.09	0.11	0.02	0.66	0.67	1.12	0.12	0.1
11	0.13	0.01	0.01	0.02	0.00	0.10	−7.30	1.12	0.12	0.1
12	0.13	0.01	0.01	0.02	0.00	0.10	−18.80	1.12	0.12	0.1
13	0.13	0.01	0.01	0.02	0.00	0.10	−36.56	1.12	0.12	0.1

K_2 is chosen as constant at 0.1 for this illustration, and the values of K_1 correspond to those in Table 4.3. "$H(T_3)$" means H from Table 4.3

LEO is subtracted from the water remaining at LL1 after sending the LLT back to the Moon (col. H in Table 4.3) and the net amount of water to arrive at the water transferred to LEO. This calculated value is compared to the guessed value. The guessed value is varied until it agrees with the calculated value. A typical spreadsheet is shown in Table 4.4 for an assumed value of K_2. Each row corresponds to the K_1 values from Table 4.3. The entire table pertains to one value of K_2. This process can be repeated for various values of K_2.

The procedure is as follows:

Step 1: For any row in Table 4.4, guess a value of water transferred to LEO and enter in col. H. This value must be less than the value of water remaining in the same row of col. H of Table 4.3.

Step 2: For this guessed value of col. H, calculate the total mass in col. C.

Step 3: Calculate the inert mass in col. D.

Step 4: Calculate the propellant mass in col. E.

Step 5: Calculate the amount of water electrolyzed in col. F and the excess oxygen in col. G.

Step 6: Repeat Steps 1–5 until the guessed value in col. H agrees with the calculated value in col. I.

Step 7: Repeat Steps 1–6 for all rows in Table 4.4.

Step 8: Repeat Steps 1–7 for various assumed values of K_2.

Finally, we estimate the requirements for sending the empty LWT back to LL1 from LEO as shown in Table 4.5. The procedure is as follows:

Table 4.5 Sample spreadsheet for calculating requirements to return the LWT from LEO to LL1

Row	Columns								
	C	D	E	F	G	H	J	K
	Total mass	Inert mass	Propellant mass	Water electrolyzed	Excess O_2	Water remaining	Rocket equation factors		
	M_t	M_i	M_p	M_{el}	M_{xs}	M_w	R	R − 1
	= D + E	= D(T_4)	= D * K	= 1.2 * E	= F − E	= M_w(T_4) − F		= J − 1
5	4.52	2.21	2.31	2.77	0.46	14.77	2.043	1.043
6	3.98	1.95	2.03	2.44	0.41	12.99	2.043	1.043
7	3.12	1.53	1.59	1.91	0.32	11.10	2.043	1.043
8	2.51	1.23	1.28	1.54	0.26	8.21	2.043	1.043
9	1.49	0.73	0.76	0.91	0.15	4.89	2.043	1.043
10	0.17	0.08	0.09	0.10	0.02	0.57	2.043	1.043
11	0.03	0.01	0.01	0.02	0.00	−7.31	2.043	1.043
12	0.03	0.01	0.01	0.02	0.00	−18.82	2.043	1.043
13	0.03	0.01	0.01	0.02	0.00	−36.57	2.043	1.043

K_2 is chosen constant at 0.1 and K_1 varies as shown in Table 4.2. "(T_4)" means taken from Table 4.4

Step 1: For each chosen value of K_2, use the value of inert mass from Table 4.4 for each row in Table 4.5 in col. D.

Step 2: Calculate the propellant mass in col. E.

Step 3: Calculate total mass in col. C.

Step 4: Calculate the propellant mass in col. E.

Step 5: Calculate the amount of water electrolyzed in col. F and the excess oxygen in col. G.

Step 6: Calculate water remaining in LEO in col. H by subtracting the water electrolyzed (col. F) from the water remaining in Table 4.4.

If the above procedures are repeated for various assumed values of K_1 and K_2, the results are as shown in Table 4.6 and 4.7.

The crux of this calculation then comes down to estimates for K_1 for the LLT and K_2 for the LWT.

For the LLT, the inert mass includes the landing structure, the spacecraft structure, the water tank, and the propulsion stage. The propulsion stage for H_2–O_2 propulsion is typically taken as roughly 12% of the propellant mass, and since the propellant mass is

Table 4.6 Mass of water transferred from lunar surface to LEO versus K_1 and K_2 if 25 mT of water are delivered to LL1

Row	Columns					
$K_1\Downarrow$ $K_2\Rightarrow$	0.10	0.20	0.25	0.30	0.35	Mined on Moon
0.10	14.77	10.98	8.71	6.14	3.18	52.78
0.14	12.99	9.66	7.67	5.41	2.80	55.38
0.18	10.85	8.06	6.40	4.52	2.34	58.52
0.22	8.21	6.10	4.84	3.41	1.78	62.38
0.26	4.88	3.63	2.88	2.03	1.04	67.25
0.30	0.57	0.42	0.32	0.24	0.12	73.57
0.34						82.12
0.38						94.31
0.42						113.14

The mass of water mined (which only depends on K_1) is also shown. All masses are in mT

Table 4.7 Fraction of water mined on the Moon that is transferred from lunar surface to LEO versus K_1 and K_2

$K_1\Downarrow$ $K_2\Rightarrow$	0.10	0.20	0.25	0.30	0.35
0.10		0.28	0.21	0.17	0.12	0.06
0.14		0.23	0.17	0.14	0.10	0.05
0.18		0.19	0.14	0.11	0.08	0.04
0.22		0.13	0.10	0.08	0.05	0.03
0.26		0.07	0.05	0.04	0.03	0.02
0.30		0.01	0.01			

Blank cells are cases where no water can be transferred

likely to be about 42% of the total mass leaving the lunar surface (from calculations), the propulsion stage is perhaps 5% of the total mass leaving the lunar surface. The water mass is likely to be about 40% of the total mass leaving the lunar surface, and if the water tanks weighs, say 10% of the water mass, the tank mass would be about 4% of the total mass. The spacecraft and landing structures are difficult to estimate. A wild guess is 12% of the total mass. Thus, we crudely estimate the value of K_1 for the lunar tanker as $0.05 + 0.04 + 0.12 \approx 0.21$. Experience shows that such estimates are always underestimates.

The LEO tanker does not require the landing system of the lunar tanker so the spacecraft mass is estimated as 7% of the total mass of this vehicle. The water transported by the LLT is about 55% of the total mass, so the water tank is estimated as 5.5% of the total mass. In addition, an aeroshell is needed that is estimated at 30% of the mass injected into LEO, which is likely to be about 90% of the mass that departs from LL1 toward LEO, so this is roughly 27% of the total mass leaving LL1. Thus K_2 for the LEO tanker is roughly estimated as 0.33 for the LEO tanker. These are of course, only rough "guesstimates."

If $K_1 \sim 0.21$ and $K_2 \sim 0.33$, only about 5% of the water extracted on the Moon is transferred to LEO. However, approximately 12% of the water launched from the Moon is transferred to LEO. It seems doubtful that this process would be economical.

4.3.2 Transfer via Lunar Orbit

In this section, we briefly compare the delivery of water from the Moon to LEO with the junction site being either LL1 or low lunar orbit (LLO). The procedure is essentially the same as that given in the previous section, except that the Δv values for each step are those that involve LLO instead of LL1 as the junction point. In doing this, the Δv values were taken from the textbook: Human Spaceflight: Mission Analysis and Design. For the LL1 case, these values are somewhat different than those used in the previous section. Table 4.8 lists these values.

Transfer via LL1 involves higher Δv for transfers to and from the lunar surface, whereas transfer via LLO involves higher Δv for transfers to and from LEO. Therefore, transfer via LL1 is expected to be more sensitive to the value of K_1 and transfer via LLO is

Table 4.8 Values of Δv (m/s) used to compare transfer via LLO and via LL1

Transfer step	Δv based on LL1	Δv based on LLO
Lunar surface to LLO or LL1	2520	1870
LLO or LL1 to lunar surface	2520	1870
LLO or LL1 to LEO	770	1310
LEO to LLO or LL1	3770	4040

Table 4.9 Fraction of mass of water mined on the Moon that is transferred via LLO and via LL1 to LEO as a function of K_1 and K_2

Based on transfer via LL1

$K_1\Downarrow/K_2\Rightarrow$	0.10	0.20	0.25	0.30
0.10	0.23	0.14	0.09	0.03
0.14	0.19	0.12	0.07	0.02
0.18	0.15	0.09	0.06	0.02
0.22	0.10	0.06	0.04	0.01
0.26	0.05	0.03	0.02	0.01
0.30				
0.34				
0.38				
0.42				

Based on transfer via LLO

$K_1\Downarrow/K_2\Rightarrow$	0.10	0.20	0.25
0.10	0.25	0.11	0.03
0.14	0.22	0.10	0.02
0.18	0.20	0.09	0.02
0.22	0.17	0.07	0.02
0.26	0.14	0.06	0.02
0.30	0.10	0.05	0.01
0.34	0.09	0.04	0.01
0.38	0.03	0.02	
0.42			

expected to be more sensitive to the value of K_2. This is illustrated by the results shown in Table 4.9.

It is probably worth noting that the cream of the crop of advanced engineering students worked on a plan to extract ice from a crater near the South Pole of the Moon (Caltech 2017). It is quite evident from the report that they did not delve into the details of surface operations, partly because as yet, nobody knows what will be encountered. The most fundamental aspect of research is the "re" in research. It is incumbent on any study to know what was done previously. Had they done a literature search they might have found the first edition of this book, published five years earlier. That might have led them to analyze the feasibility of the process, rather than just assume it was feasible.

4.4   Refueling Spacecraft with Propellants Derived from Asteroids

Asteroids might some day in the future provide a source of water that could supply hydrogen and oxygen in space that could fuel vehicles sent to Mars or elsewhere. Such a system would have a major impact on space exploration since lifting propellants from

Earth is a major contributor to the cost of space missions. However, a very significant investment in time and funds would be required to investigate this possibility fully and this futuristic scheme seems likely to be put off for many decades to come (perhaps for reasons more financial and political than technical).

Sercel (2015, 2016, 2017) developed a concept for obtaining and utilizing presumed water in asteroids [aka Near Earth Objects (NEOs)] in their native orbits. This water would be extracted by "optical mining" in which the object would be enclosed by a membrane, and concentrated solar energy (provided by inflatable reflectors) would vaporize water from the object. The water would be condensed and stored. The goal is to transfer water from the NEO to a depot in a Lunar Distant Retrograde Orbit (LDRO) which would provide hydrogen and oxygen propellants for a wide variety of space ventures. The virtue of using the LDRO as the site for the propellant depot is that the Δv requirement for transport of propellants from LDRO to LEO is only 0.7 km/s. Similarly, the Δv requirement for transport of water from the NEO to LDRO is only 0.7 km/s.

Water would be transported from the NEO to LDRO using solar thermal propulsion with water as the propellant. Solar thermal propulsion utilizes concentrated solar energy to heat a propellant to high temperatures, causing the propellant to expand out the exit nozzle imparting thrust to the system. In the original versions, hydrogen was envisaged as the propellant (Sercel 1986). In the current plan, the gaseous exudate from the asteroid (H_2O, CO_2, etc.) would be used as the propellant without processing.

Some of the water obtained from the NEO would be used up in this transfer; the remainder would be deposited at the depot at LDRO. Thus, spacecraft departing LEO for Mars (or other destinations) could receive all their hydrogen and oxygen propellants from the NEO and no propellants need be lifted from Earth. Sercel's visionary concept would revolutionize space travel. He has begun to demonstrate the required technologies with initial small levels of funding. His 2016 report provides considerable detail on how his proposed system would work. It will require a great amount of time and investment to pursue this grand concept. But the ultimate payoff could be substantial.

A related approach was proposed by Adamo and Logan (2016). One of their basic premises was:

> Water is arguably the most common volatile to be found on small bodies such as asteroids and minor moons throughout our solar system, leading to the promise of in- situ resource utilization (ISRU). With ISRU producing water for propulsion, radiation shielding, and hydration/hygiene near an interplanetary destination, mass to be transported there from Earth in support of crew return is virtually eliminated.

They proposed a reusable NTP—powered interplanetary transport utilizing water as propellant and as radiation shielding for its crew. Additional water for open-loop crew hydration and hygiene would be used.

Other approaches for obtaining water from NEOs involve mining. Zacny (2017) presented a review of various approaches for obtaining resources from asteroids. Graps et al.

(2017) provided a summary of immediate needs for identifying suitable asteroids as targets for ISRU.

Instead of attempting to recover resources directly from a NEO in its native orbit, around 2014 NASA began planning to

> … develop a first-ever robotic mission to visit a large near-Earth asteroid, collect a multi-ton boulder from its surface, and redirect it into a stable [LDRO] orbit around the Moon. Once it's there, astronauts will explore it and return with samples in the 2020s.

Fortunately, NASA decided to cancel this ill begotten concept in April 2017.

Recent NASA Plans

5

5.1 NASA ISRU Funding

Dating back to the 1990s, NASA funding for ISRU technology was minimal, consisting of at most, one or possibly two $100 K chunks per year—in a good year. In those days, I administered some of that money from JPL, supporting initial low-level work on solid oxide electrolysis of CO_2 at the University of Arizona and the Allied Signal Corporation, system analysis and sorption compressors at Lockheed-Martin Astronautics (LMA), and studies of the chemistry of methane at JPL. Jerry Sanders at JSC was also able to provide small amounts of funds for various ISRU efforts. A notable achievement was his funding to LMA to develop the cryogenic CO_2 compressor. As a result of the minimal level of funding during this period, the state of ISRU technology tended to be primitive, mainly at TRL 2–4 levels.

Around 1999, NASA funding for technology was tight and JPL conducted an internal review of its technology programs. I was at JPL and I managed a small allocation of funding for Mars ISRU. The internal JPL review decided that "ISRU is not relevant to JPL projects" and management zeroed out that small effort. Years later, in 2012, I was a retiree. But I met together with Dr. Rao Surampudi of JPL, and we proposed to JPL's internal discretionary funds that we develop a CO_2 compressor system on the grounds that any conceivable Mars ISRU system would need it. This proposal was turned down.

By a process of reasoning that would seem to defy all logic, and for reasons difficult to understand, NASA decided in the late 1990s to fund development of the Mars ISRU Precursor (MIP) to be launched as a payload on the Mars Surveyor Lander scheduled for launch in April 2001. The problems with this decision included the following:

- The state of solid oxide electrolysis technology was far too primitive to justify a flight experiment to Mars.

© Springer International Publishing AG 2018
D. Rapp, *Use of Extraterrestrial Resources for Human Space Missions to Moon or Mars*, Springer Praxis Books, https://doi.org/10.1007/978-3-319-72694-6_5

- The funding for MIP (rough guess: $10 million) was far out of balance with previous funding for research and development.
- The funds for MIP could far better have been used for advanced development of systems in the laboratory, rather than demonstrating primitive, obsolete systems on Mars.
- The experiment was risky and most likely included failure modes.
- The single cell solid oxide electrolysis unit was not easily amenable to scaling up.

As it turned out, the 2001 Mars Surveyor Lander mission was cancelled around 1999, and as a result, the laboratory precursor to the flight MIP became a museum piece. If the funds for MIP had been used to develop ISRU subsystems in the laboratory, where would the state of ISRU technology have been at the turn of the century?

NASA funding for Mars ISRU remained very low during the first decade or so of the 21st century. Technology, such as it was, was at a proof-of-concept stage. Under Griffin, funds were poured into lunar ISRU between about 2005 and 2008 but Mars ISRU remained dormant.

Suddenly, without warning or explication, on September 24, 2013, NASA HEO announced an opportunity (AO) to propose an ISRU investigation to fly on the 2020 Mars Rover, with the expectation that the cost would be about $30 M. In a sense, history repeated itself. Extraordinarily large funding was provided to demonstrate technology that was not ready for demonstration, after years of starvation. Proposals were due in 90 days. The history of NASA funding for ISRU is illustrated in Fig. 5.1. Funding of about $20 M for RESOLVE is not included.

Next, I provide some history of how JPL was able to produce the winning proposal in response to the 2013 AO. Approximately August 2013, Dr. Rao Surampudi of JPL heard a rumor that NASA was soon to issue an AO for an ISRU payload to go on the 2020 Mars Rover, which was to be more or less built like its predecessor, Curiosity. He called retiree Donald Rapp and they chatted about the possibility, and conjectured how the payload might be designed. Surampudi and Rapp had worked together on ISRU for more than 25 years, on and off, as meager funding permitted. Soon after, Rao brought in Bob Easter

Fig. 5.1 History of NASA ISRU funding

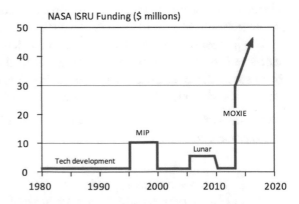

(former JPL system engineer), and Gerald Voecks (JPL expert on solid state electrolysis and fuel cells) as well as others with relevant interests and backgrounds. For about a month, a group of engineers volunteered free time to informally discuss possibilities and approaches. All of this discussion was informal, unfunded, somewhat unstructured, and definitely unofficial.

About a week went by after the AO was announced on September 24, 2013, during which JPL management debated whether to provide bid and proposal funds to support an official effort to write a proposal. Several factors figured into this. One school of thought among JPL managers was that JPL "was not in the ISRU business" and JPL is not an HEO Center; therefore it could be argued that it would be B&P funds down the drain because an HEO Center (likely JSC) would likely get the award anyway. Management also had the legitimate concern that JPL might not have proper staff to carry out such a venture.

Looking backward to late September 2013 from the vantage point of 2017, one can interpret this hesitation in several ways. At one end of the spectrum it could well be argued that JPL management was fully justified in its doubt, for ISRU was basically a new area of business for JPL, it mainly served HEO rather than space science, the spiritual leader for ISRU in the NASA community was a highly respected engineer at JSC (Jerry Sanders), HEO Centers had been actively working on ISRU for years (although almost exclusively on lunar ISRU, which is quite different from Mars ISRU), and the apparent JPL leader of the movement (Rapp) was a retiree with an uncertain reputation. Another factor was that there was no JPL upper level manager with responsibility and leadership in ISRU, writing a proposal would use up scarce B&P funds, and whoever stuck his neck out and backed a losing proposal (viewed as likely), might run into an impediment in his career path.

After the AO was released, JPL vacillated for about a week and a half until roughly October 5, 2013. Somewhere in that interim, JPL appointed a world-class proposal manager (Gail Klein) to manage production of the proposal, and at the same time, JPL set aside significant B&P funding to prepare the proposal.

At some point in early October, Mike Hecht of MIT was appointed Principal Investigator. He and Gail Klein led the production of the proposal. Gail made heroic efforts to obtain credible cost estimates, and put in inordinate hours of effort to create this proposal. Because of very constricting size limitations for the proposal, there was a great challenge in condensing down the voluminous amount of material to an acceptable size without losing the message. Hecht kept a steady hand on the steering wheel.

We were hard put to convince ourselves we could meet the deadline of submitting a proposal by the due date of December 23, 2013.

But God smiled upon the JPL-MIT Team. With great budgetary wrangling in Washington, the Congress finally shut down the Government from October 1 to October 16 in 2013. Government agencies were hamstrung, while JPL continued to operate. This slowed down production of competitive proposals by NASA Labs, but more importantly, it led to a revision of the terms of the original AO. On December 4, 2013, NASA extended the date for submission of proposals from December 23, 2013 to January 15, 2014. That gave the JPL-MIT Team three extra weeks to complete the proposal—and several team members put in a great deal of time over Christmas. With Gail Klein developing the cost estimate and Mike Hecht taking over final production of the proposal, they barely squeaked through to submit the proposal by the January deadline.

To almost everyone's surprise, the JPL-MIT Team won the competition. They began work on developing the needed technology. But at the start, nobody really appreciated the complexity of the engineering that was required, nor was anybody fully aware at first of the severe limitations and constraints imposed by placing this system in a small box on the 2020 rover. As the project evolved, it became clear that the original mandate of the AO (20 g/hr oxygen production rate) was technically impossible to fulfill, and the original cost limit of \$30 M was inadequate. As time progressed, the estimated oxygen production rate decreased and the cost increased. By 2017, oxygen production at 6–10 g/hr was the best that could be hoped for and the cost had escalated significantly. This brings up an important point. Someone, or some group at NASA decided to issue an AO calling for an oxygen production plant on Mars and required an oxygen production rate of 20 g/hr for a cost of \$30 M. Furthermore, in their optimistic projection, they assumed this project would be a springboard to a next level with an oxygen production rate 440 g/hr. Who were these people hidden behind the scenes at NASA? On what basis did they think that the ISRU technology was ready for demonstration at 20 g/hr in a package on the 2020 Mars rover for \$30 M? Where are the notes, reports, studies and documents that support this conclusion? Which ISRU experts were consulted in this regard? We will never know the answers to these questions. What we do know is (1) they were overly optimistic regarding the maturity of the technologies involved, (2) the solid oxide electrolysis system needed considerable development prior to demonstration, (3) they had little notion of the severe constraints and impediments imposed by the 2020 rover, (4) there was no way to objectively estimate the cost at the outset.

But this illustrates an important philosophical point that relates to the approach that NASA uses to develop and demonstrate new technology. NASA has two major Directorates: Space Technology and Human Exploration and Operations (ST and HEO). The evidence suggests that ST favors developing new technology from scratch in the laboratory, while HEO favors demonstrating relatively proven technology in space.

NASA is not a monolith. At the top, the NASA Administrator sets priorities. In particular, after the great success of Apollo, NASA has been trying for many decades to figure out where to send humans in space—but without resolution. Returning to the Moon, visiting an asteroid, or going to Mars are all under consideration. In the Griffin era, a firm decision was made to return to the Moon. ISRU was oriented solely to lunar applications for a number of years and funding increased. Mars ISRU was sadly neglected. When return to the Moon proved to be far more complex and expensive than had been hoped, it was shelved.

Project Constellation provides a good example of the difficulties facing NASA:

NASA formed the Constellation Program in 2005 to achieve the objectives of maintaining American presence in low-Earth orbit, returning to the Moon for purposes of establishing an outpost, and laying the foundation to explore Mars and beyond in the first half of the 21st century... The Constellation Program was conceived as a multi-decadal undertaking, with the primary goal of enabling human exploration beyond LEO... Funding for the Constellation Program was inconsistent and unreliable from its initial formulation through its cancellation. (Rhatigen 2011)

Rhatigen (2011) provided a review of Project Constellation with a set of "lessons learned". These lessons are important because they apply to any future plan to send humans to Mars or anywhere else in the solar system. If the lessons are ignored, the fate of future long-range plans will likely be unsuccessful.

Two of the many lessons learned from Constellation were as follows:

1. Robust versus Optimal Planning—The only certainty is that the funding will not match the plan driving events... The reality is that Agency flagship programs like Constellation must be robustly planned (e.g., "elastic") versus optimally planned ("inelastic"). Absent a national imperative akin to the race to the Moon, funding will not arrive as planned.
2. Schedule Creep and the Fixed Base—The Law of Diminishing Returns—Each schedule slip resulted in longer periods for which Constellation was expected to maintain the Agency's human space flight "fixed base" along with the industrial fixed base.

Now, the current vogue is to "send humans to Mars in the 2030s". Unfortunately, the Mars mission is even more complex and expensive than the lunar mission. It seems to be an inevitable fact that as Rhatigen put it: "funding will not match the plan" and "the schedule will slip".

5.2   Recent NASA ISRU Planning

The opening sentence of NASA (2015) was:

NASA is leading our nation and our world on a journey to Mars.

The report goes on to say:

Our robotic science scouts at Mars have found valuable resources for sustaining human pioneers, such as water ice just below the surface.

Yet it seems extremely unlikely that an early human mission would land at a high enough latitude to tap these resources.

The report lays down a fundamental principle:

Space exploration must be implementable in the near term with the buying power of current budgets and in the longer term with budgets commensurate with economic growth;

This is a good principle, but it is far from clear what the cost of human mission to Mars will be, and how projected budgets can provide needed funds.

NASA (2015) provides three phases for the long-term program to send humans to Mars. Phase I is *Earth Reliant* systems, such as life support, and human health and behavioral studies. Phase II is called *Proving Ground*, which includes relevant studies on habitats, and rendezvous and docking, and also include an asteroid redirect mission of uncertain value to landing humans on Mars.

The report properly acknowledges:

Entry, descent and landing (EDL) is one of our biggest challenges... A completely new approach is needed for human-scale EDL.

Yet no acknowledgement is made of the fact that development, demonstration and validation of pinpoint landing of heavy loads on Mars will require more than a decade of testing with several large-scale landings. The cost to validate this one technology might be a few tens of billions of dollars. See Adler et al. (2009).

Phase III would include an actual human mission to Mars, but it is not clear that the many challenges listed by Rapp (2015) would be met, nor is it at all clear that budgets would be adequate. NASA (2015) is an attractive, glossy, high-level public relations document but it is unconvincing technically and financially.

As we mentioned previously, NASA has two major Directorates: Space Technology and Human Exploration and Operations (ST and HEO) that share an interest in ISRU. Within those major organizations, quite a number of sub-organizations either have interest in ISRU, or have technological programs that could contribute to ISRU. There does not seem to be a unified organizational element within NASA tasked with the responsibilities to evaluate alternative ISRU approaches, and to develop ISRU systems. Lacking such a unified program in ISRU, two stalwarts within the NASA organization (Diane Linne of ST, and Jerry Sanders of HEO) joined forces to advocate to NASA management the importance of pursuing ISRU, to lead, participate in, and report on NASA studies comparing ISRU options, to keep abreast of technology advances relevant to ISRU, and to lay out plans and roadmaps for future development and maturation of ISRU. And they have done a good job.

Linne, Sanders and Taminger (2015) estimated propellant production rates needed for ISRU for lunar and Mars missions.

Linne (2016) and Sanders (2016) made presentations at a MOXIE Science Team Meeting in November 2016. Basically, these presentations were prepared to try to "sell" NASA on expanded funding for ISRU. As such, they provided optimistic visions of the future where ISRU would play a major role in enabling human missions.

Linne (2016) defined a vision of a wide assortment of potential in situ processes, not limited merely to propellant production. Linne included resource assessment (prospecting), resource acquisition, in situ construction, in situ manufacturing and in situ generation and storage of energy. Yet within that overall umbrella, she proposed focusing on a Mars O_2/CH_4 end-to-end system demonstration as primary long-term objective with the following rationale:

- Regolith acquisition and processing should/could be similar for Mars and for lunar icy soils, so we will be advancing that subsystem for both locations simultaneously.
- Atmosphere processing technology development can be structured around common components/subsystems needed for both CH_4 production from atmosphere/soil-water and O_2-only production from atmosphere.

Linne pointed out that Mars ISRU has a "ripple effect" in other exploration elements:

- MAV: propellant selection, higher rendezvous altitude (higher Δv capable with ISRU propellants).
- EDL: significantly reduces required landed mass.
- Life Support: reduce amount of ECLSS closure, reduce trash mass carried through propulsive maneuvers.
- Power: ISRU drives electrical requirements, reactant and regeneration for fuel cells for landers, rovers, and habitat backup.

Linne emphasized that ISRU will require flight demonstration missions before it will be included in the critical path, and that validation will be needed "at least 10 years before first human landed mission to ensure lessons learned can be incorporated into final design".

However, there are so many options possible that NASA could spend a lifetime evaluating each one and comparing them without ever building anything. Thus Linne said:

> The ISRU Formulation team has generated a (still incomplete) list of over 75 technical questions on more than 40 components and subsystems that need to be answered before the 'right' ISRU system will be ready for [the next] flight demo.

Throughout the presentation, Linne raised good questions, but typically there are no ready answers without further technical work.

There does not seem to be a centrally defined NASA ISRU Program. Linne (and Sanders as well) pointed out the many scattered NASA activities that could feed data or capabilities into such a NASA ISRU Program if it is ever created. Linne defined what NASA needs to do. Significant work is needed to mature ISRU technologies:

- Development and testing much closer to full-scale for human mission needs.
- Much longer operational durations.
- Much more testing outside the laboratory to validate performance under relevant environmental conditions.

- Integrate many components and subsystems into system prototypes.
- Realize synergy between ISRU and other system technologies, such as life support/fuel cell, power, surface mobility.

But will NASA do it?

Linne went on to propose an ISRU Technology Project. The end goal would be to carry out end-to-end ISRU system tests in simulated Mars environment. This would include:

- Oxygen and fuel production from atmosphere and soil resources including liquefaction and storage.
- End-to-end ISRU System integrated with power system (including excavator recharging station), thermal rejection system, autonomous control, and interfaces to life support, MAV/Lander.

An interim goal would be to carry out ISRU subsystems tests in simulated Mars environment. The idea would be to use subsystem testbeds (CO_2 acquisition and compression, O_2 production, CH_4 production, water electrolysis, icy and hydrated soil acquisition, water extraction) to evaluate multiple technology options before downselecting to the overall system.

This appears to be an intelligent approach, and NASA seemed to buy into it in early 2017. The NASA Program Executive for ISRU[1] gave a presentation in January 2017, in which the program advocated by Linne and Sanders was laid out. This involved parallel development of excavation and soil processing, and atmosphere collection and processing. Technology maturation would take place during 2017–2019, subsystems testing from 2020 and 2021, system testing from 2022 to 2024 and a Mars demonstration mission scheduled for launch in 2028. However, 2017 funding did not show up. The question remains whether NASA will continue to make presentations or whether they will actually fund the ISRU Project?

Sanders (2016) provided a companion presentation to that of Linne. He began by citing a number of NASA studies that were carried out in FY2016:

- SMD/HEOMD Mars Water ISRU Planning (M-WIP) Study (April 2016)

    - Planning for human landing site selection now includes consideration of candidate water resources in addition to the main driver: search for life.

- Evolvable Mars Campaign (EMC) Mars Atmosphere and Soil Processing Study
- ISRU Integration into Lander Ascent Vehicle and Thermal Management
- Mars "water-rich" scenario study
- ISRU and Civil Engineering Working Group (ICE WG)

    - Mars Human Landing Site Workshop (Oct. 2015)

---

[1]Nantel Suzuki, ISRU Program Executive Advanced Exploration Systems NASA HQ/HEOMD, NASA ISRU Utilization Overview, Small Bodies Assessment Group 16th Meeting, January 13, 2017.

- ISRU System Maturation Team (SMT)

  – Gap Definition/Closure (SCOREBoard) & Proving Ground Satisfaction Criteria

- ISRU Capability Leadership Team

  – Center Roles and Responsibilities
  – ISRU Facilities and Upgrade Assessment.

Evidently, there are many meetings and discussions within the NASA community relevant to ISRU.

In considering the value of various soil types for Mars ISRU, these studies considered four possible reference cases: (1) icy soil, (2) soil infused with poly-hydrated sulfates, (3) clay, and (4) typical regolith. Icy soil is obviously the best source of indigenous water, but it is limited to higher latitudes. Soil infused with poly-hydrated sulfates can yield 8% water at 100–150 °C, and this is clearly superior to clay or regolith as a souce of water.

A critical issue for Mars ISRU is whether to be content with an oxygen-only ISRU system and bring methane from Earth. This is discussed in Sect. 3.1.2. While Sanders and Linne seem to be driving NASA toward water-based ISRU, the arguments pro and con appear to be fairly balanced.

Summary and Conclusions

6

Since the earliest expeditions of humans into space, visionaries have contemplated the possibility that extraterrestrial resources could be developed and civilization could eventually move into space. (Most recently, Elon Musk said "We'll create a city on Mars with a million inhabitants".)[1] As time went by, visionaries looked beyond the near-term and imagined the transfer of the industrial revolution and the electronic revolution to planetary bodies. ISRU visionaries know no bounds. Imaginative proposals abound for all sorts of futuristic systems.[2]

NASA is not a monolithic organization. Two cultures that exist side-by-side within NASA are the project implementers and the advanced technologists. The project implementers desire to carry out space exploration projects as safely, reliably and effectively as possible, which often implies using proven technology rather than newly emerging technology. Their unstated motto is "better is the enemy of good enough". The advanced technologists seek to provide improved performance with new concepts but these emergent concepts are not typically funded to maturity, nor are they typically fully vetted for reliability. As a result, there is a certain tension between the two cultures and lack of transfer of new technology from the laboratory to the projects has been and remains a problem for NASA. At the same time, technologists seem to have difficulty developing (or getting funds to develop) technology to greater maturity, leaving many technologies unready for direct application in flight projects.

Imbedded within NASA is a small cadre of ISRU technology enthusiasts who are constantly seeking support from the greater NASA for further development of ISRU through advocacy, presentations and analyses. These are amplified by a number of devotees outside of NASA. Since the 1990s, the enthusiasts developed elaborate plans for development of ISRU technology that include the more mundane elements (propellant

---

[1]http://www.telegraph.co.uk/science/2017/06/21/elon-musk-create-city-mars-million-inhabitants/.
[2]http://www.isruinfo.com/.

© Springer International Publishing AG 2018
D. Rapp, *Use of Extraterrestrial Resources for Human Space Missions to Moon or Mars*, Springer Praxis Books, https://doi.org/10.1007/978-3-319-72694-6_6

production, life support) as well as more ambitious elements (e.g. "in-situ manufacturing and assembly of complex parts and equipment", "in-situ fabrication and repair").

Traditionally, about half of the NASA budget has been addressed (in one form or another) to humans in space, about one quarter for science, and one quarter for smaller elements. However, after the remarkable achievements of the Apollo Program, NASA has not been able to establish a viable role for humans in space. After completion of the amazing Apollo Program, NASA elected to develop a generic reusable access to space in the form of the Space Shuttle (SS). The Shuttle accomplished some useful tasks, but by and large, the payoff was far less than proportional to the large investment. The Shuttle became problematic with age, and reached the point where its later flights seemed to have one major mission objective, and that was to do enough repairs in space to assure a safe landing on return! The Shuttle was not a springboard for venturing further into deeper space. The International Space Station (ISS) has proven to be a very expensive boondoggle and we still don't know exactly what to do with it. It is a poor place to mount telescopes and has little other utility other than use as a testbed for effects of space on astronauts.

When Mr. Griffin took over the helm of NASA in 2005, he attempted to push NASA toward a specific destination: the Moon. He attempted to phase out the SS and ISS to create budget room for his newly proposed Exploration Program that would be aimed almost exclusively at a return to the Moon. He met with storm of resistance from certain Congressmen who will not stand for any interruption in "America's access to space". Some Congressmen preferred to prolong operation of the SS so it could service the ISS (apparently, merely to follow through with our international partners) when there was no longer a need for the SS, and there never was much need for ISS. Faced with resistance to his desire to phase out SS and ISS, Dr. Griffin decided to tap other programs within NASA for funds. As a result, NASA budgets were thrown into disarray and NASA activity was focused on how to return humans to the Moon affordably using a minimum of new technology. NASA technology and exploration programs were severely cut back and all previous plans were made instantly obsolete. This continued a long-term trend of on again—off again support for NASA technology programs. Several hasty studies were conducted by the inner circle supporting Dr. Griffin, to evaluate the feasibility, requirements and cost of returning humans to the Moon. As one might expect, the conclusions were optimistic. The push toward a lunar return continued, but it wasn't until a few years later that workshops were held to try to answer the question why we should return to the Moon and what were the payoffs? These workshops did not provide good support for a lunar return. As the years went by, the perceived cost and difficulty of returning humans to the Moon went up, and the rationale for why NASA should return to the Moon remained nebulous. Congress balked at the cost. By 2012, the impetus to return to the Moon drifted off to the backburner.

Nevertheless, during the several years after 2005, when the lunar initiative was in its peak period, NASA made significant new investments in ISRU technology—but only for the Moon. Essentially no funds were available for Mars ISRU. Thus, in the post-2005 period, lunar ISRU received significant new funding from NASA but Mars was ignored. The managers of the NASA ISRU technology program could not possibly benefit from a system study that was likely to show that lunar ISRU is impractical. That would leave them with no basis for the program they were appointed and anointed to lead. Imagine yourself barely getting along for years running NASA's ISRU program on a pittance and suddenly, large funding drops in your lap—but only for lunar ISRU. What would you do? I would take the money and run. That is what they did. By simply repeating the mantra that ISRU reduces IMLEO, they avoided detailed discussion of costs and feasibility of lunar ISRU.

Lunar resources are discussed in Sect. 2.1. These include silicates in regolith, FeO in regolith, imbedded atoms in regolith from solar wind, and water ice in regolith pores in permanently shadowed craters near the poles. As we showed in Sect. 2.2, the processes for utilizing these resources require huge amounts of regolith, require large amounts of power that might not be available, and involve operations that appear to be impractical. Obtaining oxygen from silicates requires extreme reaction conditions (2600 K) that suggest the process is impractical. Obtaining oxygen from FeO is more feasible, but it still requires processing huge amounts of regolith and high power levels, and questions remain about the feasibility of autonomously processing solids in a batch mode. Imbedded atoms from solar wind are very sparse and require processing huge amounts of regolith with high power requirements. Obtaining water ice in regolith pores in permanently shadowed craters requires heavy machinery operating in a cold, dark environment with high power requirements and no solar energy.

Furthermore, the main product of these processes is oxygen, and there probably won't be much demand for oxygen generated by ISRU on the Moon because according to project plans, oxygen would probably not be used as an ascent propellant, ascent propellants would have to be brought from Earth anyway if abort-to-orbit is required, and an ECLSS will reduce the moderate oxygen requirement for breathing. Furthermore, even if oxygen were to be used as an ascent propellant, the leverage inherent in lunar propellant production is far less than that for Mars propellant production.

Aside from the technical difficulty in implementing lunar ISRU, and the lack of project needs for ISRU products, there is also the question of affordability. The affordability of water extraction from shadowed ice in craters depends on the cost of prospecting to locate the best sources of water ice, as well as the operational missions to acquire water and deliver it. Preliminary analysis suggests that the cost is likely to be too high to be affordable.

By contrast, Mars ISRU has attractive possibilities. The RWGS process requires a good deal of work; yet it might become practical. The MOXIE Project has developed solid oxide electrolysis to an advanced state of maturity. The S/E process is entirely feasible but requires a source of hydrogen. While the cost of prospecting for near surface H_2O on Mars will be expensive, it is likely to be worth the investment. In the process of doing this, the opportunity to do some valuable Mars science studies will present itself. The processes: SOXE, RWGS and S/E have the potential to yield important reductions in IMLEO for human missions to Mars by providing ascent propellants, while the S/E process with indigenous water also provides additional safety and reliability in the form of backup for an ECLSS system. The S/E system is fully proven; all it needs is a supply of hydrogen.

Use of ISRU to generate ascent propellants on Mars provides two basic benefits. One is that it eliminates the need to transport tens of tons of cryogenic ascent propellants from Earth. The other is that it allows the orbiter to remain in an elliptical orbit. This orbit requires extra ascent propellants (compared to a circular orbit) but these propellants can be produced by ISRU. The Δv for insertion into this orbit, and for departure from this orbit is lower than that for insertion and departure from a circular orbit. This reduces the amount of propellant that must be brought from Earth for orbit insertion and departure of the ERV. Overall, the reduction in IMLEO when ISRU is employed is of the order of 300–400 mT, and perhaps surprisingly, a significant part of this derives from the lower Δv requirements for insertion into and departure from the elliptical orbit (as opposed to a circular orbit). There are also significant benefits in regard to reliability of life support if indigenous water is obtained on Mars.

As we have discussed, NASA seems to have two major thrusts. One is the search for extraterrestrial life in the solar system and beyond using robotic and telescopic technologies. The other is to send humans into space, but here there is no central theme or goal. In fact NASA has not yet figured out why or where they want to send humans into space, except that their budgets would be much smaller if they didn't. So, there is a bigger issue at hand than ISRU: what aims should the NASA space program adopt, and what role (if any) should ISRU play in this?

In October 2004, more than 130 terrestrial and planetary scientists met in Jackson Hole, WY, to discuss early Mars. The search for life on Mars was a central theme in their report. The word "life" occurred 119 times in their 26-page report, an average of almost five times per page. The Introduction to that report said: "Perhaps the single greatest reason scientists find this early period of Martian geologic history so compelling is that its dynamic character may have given rise to conditions suitable for the development of life, the creation of habitable environments for that life to colonize, and the subsequent preservation of evidence of those early environments in the geologic record." One of the three "top science questions related to early Mars" was stated to be "Did life arise on early Mars?" Later in the report it says: "The question of Martian life embodies essentially three basic aspects. The first involves the possibility that Mars may have sustained an independent origin of life. The second involves the potential for life to have developed on one planet and be subsequently transferred to another by impact ejection and gravitational capture (i.e., panspermia). The third focuses on the potential for Mars to have sustained and evolved life, following its initial appearance." The report then admits that: "How life begins anywhere remains a fundamental unsolved mystery" and it further admits that: "The proximity of Mars and Earth creates an ambiguity as to whether Earth and Mars hosted truly independent origins of life. Meteoritic impacts like those that delivered Martian meteorites to Earth might also have exchanged microorganisms between the two planets. Impact events were even more frequent and substantial in the distant geologic past, including a period of time after which life began on Earth. Thus we cannot be sure whether the discovery of life on Mars would necessarily constitute the discovery of a truly independent origin of life."

According to the Mars Exploration Program:

The defining question for Mars exploration is: Life on Mars?

Similarly, the main basis of NASA Exploration on other bodies in the solar system and beyond is the search for life.

The emphasis on the search for life in the NASA community has swayed a number of otherwise competent and prominent scientists to develop programs, papers, and reports to analyze, hypothesize and imagine the possibility of life on other planetary bodies, with prime emphasis on Mars—and these occasional musings have been blown out of proportion by the press. A not-so-subtle pressure weighs on Mars scientists to find implications for life in studies of Mars.

One of the great, unsolved puzzles in science is how life began on Earth. The prevailing view amongst scientists today seems to be that life forms fairly easily with high probability on a planet, given that you start with a temperate climate, liquid water, carbon dioxide, and perhaps ammonia, hydrogen and other basic chemicals, and electrical discharges (lightning) to break up the molecules to form free radicals that can react with one another.

The fact that life arose on Earth comparatively early in the history of the Earth seems to be the foundation of the widely believed argument that life forms easily and with high probability—an argument that has no basis that I can discern. First of all, we don't know if life "began" on Earth or was transferred to Earth from another body. Secondly, we don't understand the process by which life was formed, so how can we be sure that the relatively early emergence of life on Earth is indicative of anything? There is no evidence or logic to suggest that if life arose say, 3 billion years after the formation of the Earth (rather than 1 billion), this would require that the probability of life forming is lower than if it formed in 1 billion years. And even if this argument held, it would only be a factor of three, whereas the innate probability of forming life must be a very large negative exponential.

Considering the complexity of life—even the simplest bacterium requires perhaps more than a thousand complex organic enzymes in order to function—the probability seems extremely small that starting with simple inorganic molecules, life would evolve spontaneously. The problem for all of the explanations of the origin of life from inanimate matter is that none of them hold up to even cursory scrutiny. The one thing that seems most likely is that the innate probability is very small, and given say, 1,000,000 planets in the universe with a climate that could theoretically support life, it is possible that only an extremely rare and fortuitous conflux of events led to the formation of life on one planet (or possibly a few).

Hence the entire NASA exploration program to search for extraterrestrial life is a very long shot with the odds exponentially favoring failure. On the other hand, exploration of planets provides us with knowledge and understanding of the solar system that has some value. Whether it is worth the investment, remains a matter of opinion. We spend more than three times as much on NASA than we do on cancer research.

While the basis for robotic exploration of the solar system in the search for life is weak, the basis for sending humans into space is weaker still. The MIT Space, Policy, and Society Research Group argued[3]:

> We define primary objectives of human spaceflight as those that can only be accomplished through the physical presence of human beings, have benefits that exceed the opportunity costs, and are worthy of significant risk to human life. These include exploration, national pride, and international prestige and leadership. Human spaceflight achieves its goals and appeals to the broadest number of people when it represents an expansion of human experience. Secondary objectives have benefits that accrue from human presence in space but do not by themselves justify the cost and the risk. These include science, economic development, new technologies, and education.

It seems likely that national pride and international leadership was an important motivating factor in the Sputnik era when the U.S. was in competition with the Soviet Union, but that seems to have faded quite a bit with the passage of many years after Apollo with little to show for it. The requirement that human spaceflight engage in activities that "can only be accomplished through the physical presence of human beings" is somewhat of a tautology. The unstated question is whether the things that only human beings can do are worth investing in and whether the cost, complexity and risk associated with sending humans into space outweighs their greater flexibility to deal with the unforeseen.

The amazing thing is that the MIT Space, Policy, and Society Research Group recommended that NASA "continue to fly the Space Shuttle" and "develop a broad, funded plan to utilize the ISS through 2020 to support the primary objectives of exploration". By late 2008, the Space Shuttle had become more of a liability than an asset, and the ISS was never good for very much.

The MIT group also recommended: "A new policy should direct the balance between the Moon, Mars, and other points of interest in future explorations". This is not as simple as it sounds. Michael Griffin (NASA Administrator in 2007) said:

> Most of the next 15 years will be spent re-creating capabilities we once had, and discarded. The next lunar transportation system will offer somewhat more capability than Apollo. But in all fairness, the capabilities inherent in [the vehicles of the 2010–2020 period] are not qualitatively different than those of Apollo, and certainly are not beyond the evolutionary capability of Apollo-era systems, had we taken that course. But we did not, and the path back out into the solar systems begins, inevitably, with a lengthy effort to develop systems comparable to those we once owned. It will cost us about $85 billion in Fiscal 2000 currency to get to the seventh lunar landing by 2020.

It is now believed that the cost would be considerably higher than the above estimate, but the rest of the statement is correct.

Griffin went on to say:

---

[3]http://web.mit.edu/mitsps/MITFutureofHumanSpaceflight.pdf.

The above assessment is, for many, a bitter pill to swallow. Not only is it depressing for advocates of human exploration to face the fact that so many years will be spent plowing old ground, but there is also the question of why it will take so long. Again, the answer is captured in the funding profile.

Griffin estimated the cost of a human campaign to Mars

... at \$156 billion in Fiscal 2000 dollars. If \$4.8 billion/year is available in the human spaceflight account, then the Mars mission development cycle will require about 15 years. Thus, if we begin development work in 2021, we will be able to touch down on the Martian surface in about 2037, with follow-on missions every 26 months thereafter for the next two decades.

It seems possible on intuitive grounds that the relative cost of campaigns to LEO, the Moon, and Mars are in the relative ratios: 1:10:100. If a lunar campaign costs well over \$100 billion, it seems likely that a Mars campaign might cost a trillion dollars. But that is about one year's federal budget deficit, so why not?

We have seen that there is really no fundamental reason to send humans to remote places in the solar system, certainly not in the early to mid-21st century. Nevertheless, there is enough inertia in the government system that humans in space is highly likely to continue, so it is pointless to argue whether or not it should be continued. Instead, the issue comes down to what should NASA do, given that roughly \$10 billion a year will be thrust upon it for human space flight and operations?

The problem is this. NASA has already done most of what humans can do in Earth orbit. NASA has landed humans on the Moon—a dull, lifeless, uninteresting body, with precious few resources. The next step is to land on Mars. The great disparity between the time required in traveling to Mars (6–9 months) vs. the time required to travel to the Moon (~ 3 days) results in a huge difference in requirements for transporting the crew in the two cases. These differences include differences in mass of life support consumables, requirements for longevity and durability of ECLSS systems, radiation exposure, low gravity exposure, and habitat volumetric space and facilities. When the entire mission is taken into account, including transit to Mars, surface stay, and return, the differences between Mars missions and lunar missions are very significant. A lunar mission can be aborted at almost any time in the mission and return to Earth can be achieved in a few days. In a Mars mission, once the spacecraft makes significant headway toward Mars, there is no turning back (the Earth has moved away around the Sun). The total mission must be carried out from start to finish (about 200 days traverse to Mars; 550 days on the surface; and 200 more days for the return flight). As a result, the level of reliability must be incredibly high for human missions to Mars. There are also significant differences in Δv requirements for various transits, although these can be partly mitigated by aero-assist technologies at Mars, and possibly advanced propulsion. The large amounts of propellants, the large aeroshell, and the large habitat and ascent vehicle that must be sent to Mars dictate that a very large amount of mass must be launched to LEO to support a human mission to Mars. NASA has consistently optimistically estimated this mass. Mars launch opportunities are spaced at 26-month intervals with significant variance in launch and return characteristics at each opportunity. Significant Δv variations from opportunity to opportunity require that a design be adopted for the worst case. In case of an unforeseen problem, a rescue mission cannot be launched until the next opportunity. As a result, excessive use of the lunar paradigm for Mars is inappropriate. ISRU on Mars has much greater impact on the mission than lunar ISRU has on lunar missions. Mars ISRU processes are inherently more feasible and more efficient than lunar ISRU processes.

We therefore must conclude that the only appropriate target for ISRU to enhance human exploration of the solar system in the 21st century is in establishing a human outpost on Mars. Yet, the technical difficulties and economic challenges involved in a human mission to Mars are significant and have been repeatedly optimistically estimated by NASA as well as private enthusiasts. The only feasible way to approach this is to unite the robotics and human exploration programs into a single integrated effort to develop the needed capabilities in advance of the actual mission. Even so, it will take a long time and the cost will be high.

The first step in the process is to locate a site or sites to develop as a hub for eventual establishment of a human outpost. The Mars robotic exploration program must cease and desist pursuing its endless quest to discover signs of putative early life on Mars via minute concentrations of carbon. Instead, the goal of the Mars robotics program should be to locate the best near-surface H_2O resources in regions of Mars suitable for a landing. This site will eventually provide water, breathing air, ascent propellants and propellants for surface mobility on Mars. The tantalizing observation by Mars Odyssey from orbit was that three large near-equatorial regions encompassing millions of square km contain perhaps 8% H_2O in the upper meter of regolith. If this were accessible, it would provide an extensive source of hydrogen. Thus, the main problem for this form of ISRU on Mars is not process development, but rather, prospecting for near-surface H_2O.

What is needed is a campaign beginning with long-range, near-surface observations with a neutron/gamma ray spectrometer (or other instruments to detect hydrogen) in the regions of Mars identified from orbit as endowed with near-surface H_2O. The range of a rover is limited, and NASA must develop the capability for intermediate range exploration: much greater range than a rover (10–100s of km), but capable of much more precise observations than is possible from orbit. This might involve balloons, airplanes or gliders, network landers, or possibly an orbiter that dips down to low altitudes for brief periods. A dipping orbiter could observe the surface from ~ 70 km instead of 400 km, thus reducing the pixel size from 300 km to about 100 km. A network of mini-landers could examine a number of spots within a desirable region. Balloons, airplanes and gliders could make close observations over tens to 100s of km. None of these technologies seem to be presently high on NASA's priority list. Over the past 20 years, many proposals were submitted to NASA for innovative concepts for intermediate range exploration. Mars balloons, gliders and airplanes were proposed to the Mars Scout Program but these were all rejected in favor of more mundane missions that were viewed as less risky. In 2003, JPL submitted a proposal to develop non-propulsive foam landing systems for deploying networks of instruments on Mars. NASA had no interest in this technology. Using several of these approaches, the object is to locate to within rover range (several km) within the 300 km pixel of Mars Odyssey, optimal locations with H_2O content $\sim 8\%$ in the top 1 m.

The second stage of prospecting exploration is to send in a few rovers to explore the optimal locations first-hand. These rovers would pinpoint the optimal H_2O deposits both horizontally and vertically to the nearest meter. They would also place a beacon to assist in pinpoint landing for the next step.

The third stage of prospecting exploration is to send in a rover equipped with a subsurface excavation technology to pull out samples for ground-truth to verify the rover measurements, and determine the form of H_2O, as well as the soil characteristics and accessibility of subsurface H_2O. This will require pinpoint landing capability.

The fourth stage is to send in heavy equipment to work the surface, extract and purify water, and store water for the future. This could temporarily be solar powered.

These four stages are likely to require more than a decade and several billion dollars. Solar power and RTGs would be used for the first four stages.

While all of this robotic exploration in search of optimal near-surface H_2O on Mars is in progress for two decades, the human exploration arm of NASA would be busy, not sending humans into LEO to accomplish very little of significance, but rather preparing subsystems for eventual use in a human mission to Mars. This would include:

- Development of a heavy lift launch vehicle
- Development and Demonstration at Mars of large-scale EDL
- Development of a Mars Habitat
- Development of a Nuclear Power System
- Development of a long-life resilient ECLSS
- Development of suitable radiation shields
- Development of artificial gravity systems
- Development of propulsion systems compatible with ISRU
- Development of a prototype ascent vehicle

In the fifth stage, possibly 20 years into the campaign, major deliveries would be made to the site:

A full-scale nuclear reactor power system (It would be stored behind a regolith berm to eliminate line-of-sight radiation exposure, and power lines would carry power to a central power distribution station.).

- Full scale heavy equipment to work the surface
- Full scale ISRU processing system
- An ascent vehicle with cryogenic propellant storage

As water accumulates, it can gradually be converted to methane and oxygen and stored in an ascent vehicle. Storage of these cryogenic propellants requires electrical power.

The 6th stage in the process would be to land a Habitat on Mars and keep it functioning for a period of at least two years.

All of the landed systems up to this point are regarded as prototypes and are used as redundant backup for the eventual outpost. After two decades into the campaign, NASA could contemplate beginning the mission sequence shown in Fig. 1.1, assuming the various vehicles and systems that are required were developed and demonstrated.

All of the above pertain to the next few decades. In the longer run, obtaining water from NEOs might revolutionize space exploration and travel (see Sect. 4.4).

Appendix A
Transporting Hydrogen to the Moon or Mars and Storing It There

This is a digest of a report written in 2005 by D. Rapp.

A.1 Storage as High Pressure Gas At ~ Room Temperature

The problem with storage as a high-pressure gas at ordinary temperatures is that the pressure must be very high and therefore the tank mass is high. About 95% of the total mass is tankage and about 5% is hydrogen. Historical performance of hydrogen gas storage tanks is as follows:

Pre-1980: Type I steel tanks ($\sim$1.5% H_2 by weight)
1980–1987: Type II hoop-wrapped tanks ($\sim$2.3% H_2 by weight)
1987–1993: Type III fully-wrapped Al tanks ($\sim$3% H_2 by weight)
1993–1998: Type IV all-composite tanks ($\sim$4.5% H_2 by weight)
2000–2003: Advanced composite tanks ($\sim$7% H_2 by weight)

The overall system including mountings, in-tank regulator and plumbing would provide a lower effective [percentage of hydrogen]. For example a 7% hydrogen tank might provide 5% hydrogen for the entire storage system. Tanks at 5000 psi store a higher percentage of H_2 by weight, whereas storage at 10,000 psi reduces the storage volume.

A.2 Storage as a Cryogenic Liquid

A.2.1 Mass Factors

A rough rule of thumb is that one can store liquid hydrogen in a tank with a mass distribution of roughly 20% tank mass, 75% usable hydrogen mass, and 5% ullage and residual hydrogen. This does not include other mass effects such as the need for shields, or implications for spacecraft structure induced by inclusion of cryogenic tanks. The problem with cryogenic storage, unlike storage as a high-pressure gas, is not mass; it is the problem of heat leaks and boil-off.

The question of viability for various space applications of various durations will depend on the rate of heat leak into the tank, resulting in boil-off over the duration of the mission. Alternatively, one could possibly use active refrigeration to remove heat from the tank at the rate that heat leaks in, resulting in "zero boil-off" (ZBO). However, the mass and power requirements for ZBO, and issues related to reliability are complex.

© Springer International Publishing AG 2018
D. Rapp, *Use of Extraterrestrial Resources for Human Space Missions to Moon or Mars*, Springer Praxis Books, https://doi.org/10.1007/978-3-319-72694-6

As in the case of high-pressure storage, there are many Internet sites that deal in one way or another with storage of liquid hydrogen, including many DOE sites, but the word "mass" rarely appears in any of them.

A.2.2 Rate of Boil-Off from Hydrogen Tanks in the Vacuum of Space

The obvious choice for insulation in space is multi-layer insulation (MLI). However, MLI is ineffective in an atmosphere and it would be necessary to undercoat the MLI with an insulation that is effective in an atmosphere such as foam or evacuated micro-spheres for use during launch operations. This undercoat of insulation that can function in an atmosphere would be necessary for storage on Mars where there is an atmosphere, since the MLI will not provide very much insulation capability on Mars. However, heat transfer through any insulation designed for an atmosphere will be faster than it would be through evacuated MLI.

Boil-off on the launch pad is typically controlled by foam insulation. This boil-off has been estimated at 1.2% per hour. That is why launch vehicles are "topped off" as near to launch time as possible.

Boil-off in space is controlled by MLI insulation. Guernsey et al. (2005) recommended use of the following equation for the heat flow (W/m^2) through N-layers of MLI, where T_H is the absolute external temperature and T_L is the absolute temperature of the liquid hydrogen in the tank.

$$Q_{MLI}/A = \{1.8/N\}\{1.022 \times 10^{-4} \times [(T_H + T_L)/2][T_H - T_L]$$
$$+ 1.67 \times 10^{-11}(T_H^{4.67} - T_L^{4.67})\}$$

For example, if $N = 40$, $T_H = 302$, and $T_L = 20$, we calculate $Q_{MLI}/A = 0.5\,W/m^2$.

The relationship between the heat leak rate Q_{MLI}/A (W/m^2) and the hydrogen boil-off rate can be derived by using the heat of vaporization (446 kJ/kg). The boil-off in kg/day per m^2 of tank surface is

$$(Q_{MLI}/A) \times 24 \times 3600/446,000 = 0.194(Q_{MLI}/A)\,kg/day \text{ per } m^2 \text{of tank surface.}$$

For a tank of diameter 3.3-m with a surface area of 34 m^2, this would be 6.6 (Q_{MLI}/A) kg/day for the entire tank.

This tank holds $(4/3)(\pi)(3.3/2)^3(70) = 1917$ kg of hydrogen.

Therefore the percentage loss of hydrogen is

$$6.6 \times 100(Q_{MLI}/A)/1917 = 0.344(Q_{MLI}/A)(\% \text{ per day})$$
$$\text{or } 10.3\,(Q_{MLI}/A)(\% \text{ per month}).$$

For the estimate from the above formula, (Q_{MLI}/A) $\sim$ 0.5 W/m^2, and this implies a loss rate of $\sim$0.17% per day or $\sim$5% per month. This heat load could presumably be reduced by adding more layers of MLI. This is an idealized rough estimate of the heat leak into a hydrogen tank surrounded by a vacuum environment at 302 K. In actuality, for any application, one should take into account the total environment of the hydrogen tanks, including their view to space, their view to other spacecraft elements, possible use of shades for shielding the tank from the Sun and planetary bodies, conduction though the support structure (struts) and obstructions that block the view to space, such as struts, thrusters, other tanks, and miscellaneous objects. A tank that would be earmarked for landing on Mars would undoubtedly be mounted inside an aeroshell en route to Mars. Configurations that are planned for landing on the Moon or Mars would likely use a number of smaller tanks for packing density, rather than one large tank, and this would increase the heat load.

A comparison of predictions for boil-off rate is given in Fig. A.1. The NASA/JSC estimate[1] was generated in 2004. The heat transfer formula was given previously. The corrections for seams and penetrations were recommended by Guernsey et al. (2005). Augustynowicz et al. (1999) would have predicted heat leak rates about double those of the upper curve.

MLI degradation due to seams and penetrations and for conduction through supports and pipes should be taken into account. The MLI degradation factor is the ratio of heat leak due to seams and penetrations to the heat transfer rate through the MLI blanket. The conduction factor is the ratio of non-insulation heat transfer to heat transfer rates through the insulation system. Haberbusch et al. (2004) recommended use of an MLI factor 1.74 and a conduction factor of 0.14. Guernsey et al. (2005) discussed seams, penetrations and conduction through supports, and recommends adding 20% to the calculated heat leak and then adding a 50% margin on top of that to cover uncertainties. Augustynowicz et al. (1999) said "Actual thermal performance of standard multilayer insulation (MLI) is several times worse than laboratory performance and often 10 times worse than ideal performance." A comparison of estimates is given in Fig. A.1.

Handling and installation of MLI is tricky. If the MLI is compressed, it will "short out" layers and increase conductivity. Augustynowicz et al. (1999) presented data on heat leaks through MLI into cryogenic tanks. Their data presented in Fig. A.2 are in terms of the equivalent effective thermal conductivity, which must be multiplied by the temperature differential, and divided by the thickness to obtain the heat flow per unit area. For 40-layer MLI in a high vacuum, k approaches about 0.0006 W/m-K. This MLI has a thickness of about 0.22 m. For a temperature difference of 280 K, the heat flow per unit area is $0.0006 \times 280/0.22 = 0.8$ W/m^2. This would lead to a boil-off rate of $\sim 8\%$ per month.

Any system that stores liquid cryogen in space will inevitably have some heat leaks that cause vaporization of the cryogen, resulting in pressurization of the tank if the excess vapor is not vented. Venting without loss of liquid cryogen is a tricky business in zero-g.

---

[1]Lunar Architecture Focused Trade Study Final Report", 22 October 2004, NASA Report ESMD-RQ-0005.

Fig. A.1 Comparison of predictions for boil-off rate

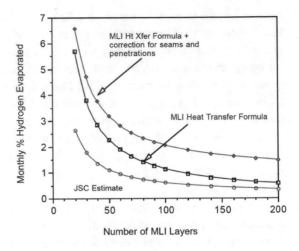

Fig. A.2 Performance of insulation versus pressure (torr). The MLI is superior at high vacuums whereas the aerogel is superior at Mars pressures (4–8 torr)

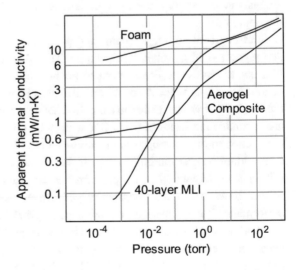

Nevertheless, it can be done. Thermodynamic venting methods for venting without loss of liquid are under development. Space telescopes that utilize cryogenic cooling typically employ a porous plug to allow venting of helium without loss of liquid helium but this may not be applicable to hydrogen. The viability of liquid hydrogen storage for any space application is critically dependent on the rate of heat leak into the tank. If venting of boil-off is used, the rate of heat leak will determine how over-sized the initial tank must be to provide the required mass of hydrogen after the required storage duration. There are indications that hydrogen can be stored so that the initial amount of useable hydrogen accounts for 75% of the initial total mass, and based on this, we can estimate the required initial total mass of storage if venting amounts to X % per month for M months. In the following, subscripts "I" and "F" correspond to initial and final, subscripts "H" and "T"

refer to hydrogen and tank, respectively, and "TOT" refers to total (hydrogen + tank). Thus:

$$M_{TOT,I} = M_{H,I} + M_{T,I} = 1.33 M_{H,I}$$
$$M_{T,F} = M_{T,I}$$
$$M_{H,F} = M_{H,I}(1 - MX/100)$$
$$M_{TOT,I} = 1.33 M_{H,F}/(1 - MX/100)$$

For example, if $M = 7$ months and $X = 7\%$, the initial total mass is 2.61 times the final delivered hydrogen mass, so the final hydrogen mass is 38% of the initial total mass. Clearly, when (MX) approaches 100, the required initial mass blows up. Using the estimates of Plachta and Kittel (2003) we can derive the following inferences regarding what the initial mass must be in order to provide a given amount of hydrogen some time later, for any boil-off rate. For any tank of radius R initially filled with liquid hydrogen, the tank mass is

$$M_T = 6.3 \times A = 6.3 \times 4\pi R^2 (\text{kg})$$

and the initial hydrogen mass is

$$M_I = 70\,V = 70 \times (4/3)(\pi)R^3 (\text{kg})$$

The initial total mass is

$$M_{TOT} = 6.3 \times 4\pi R^2 + 70 \times (4/3)(\pi)R^3 (\text{kg})$$

The rate of hydrogen boil-off in the tank is proportional to the tank surface area, $A = 4\pi R^2$. Hence the final mass of hydrogen in the tank after storage for (N) days is

$$M_F = M_I - KAN$$

where K is a constant that depends on the rate of heat transfer into the tank and the heat of vaporization of hydrogen. **Plachta and Kittel** estimated that to provide 1250 kg of hydrogen after 62 days, the required initial total mass of the tank plus initial charge of hydrogen is ~ 1700 kg. Using $M_{TOT} = 1700$, we find $R = 1.71$ m, $M_T = 232$ kg, $M_I = 1468$ kg and $A = 36.75$ m^2. Thus we conclude that according to the model of Plachta and Kittel:

$$K = (M_I - M_F)/(AN) = (1468 - 1250)/(36.75 \times 62) = 0.096 (\text{kg/m}^2\text{-days}).$$

Thus the heat leak implied by Plachta and Kittel is estimated to be $0.096 \times 446{,}000$ J/kg/(24 × 3600 s/day) = 0.50 W per m^2 of surface area.

This is considerably less than the value 0.9 W/m^2 derived previously for 40 layers of MLI. Had we used 0.9 W/m^2, K would have been 0.173 (kg/m^2-days). For a nine-month passage to Mars, $N = 270$ days. In this case,

$$M_I = M_F + 270\,K\,A$$

But from the expression for M_I, we can convert A to M_I using:

$$A = 4\,\pi\,R^2 = 4\,\pi\,[M_I/(70 \times (4/3)(\pi)]2/3$$

Hence we can solve for M_I from M_F and N by successive approximations. Haberbusch et al. modeled hydrogen storage with thick MLI blankets up to ~ 300 layers. However, they did not present results for passive storage, and only considered zero boil-off systems.

In summary, a hydrogen tank insulated with MLI in a vacuum will acquire heat from surroundings at a rate that depends on the temperature distribution of its surroundings and the number of layers of MLI. There are various estimates of the heat leak rate and therefore the boil-off rate, and it is difficult to pinpoint the boil-off rate. For a tank surrounded by an environment at room temperature, with about 40–50 layers of MLI, the estimated heat leak is somewhere in the range 0.4–0.9 W per m^2 of surface, and this will boil off at rates that might be in the range ~ 4 to 9% per month. Thicker blankets of MLI should reduce the boil-off but penetrations and seams will still play their role.

A.2.3 Rate of Boil-Off from Hydrogen Tanks on Mars

The atmospheric pressure on Mars varies from about 4 to 8 torr. At these pressures, MLI is far less effective than it would be in a good vacuum. Augustynowicz et al. provided the data shown in Fig. A.2 for various types of insulation. It is noteworthy that all the insulations perform better (smaller thermal conductivity, k) at lower pressures. The data presented in the this figure are in terms of the equivalent effective thermal conductivity, which must be multiplied by the temperature differential, and divided by the thickness to obtain the heat flow per unit area. Using either foam or the best aerogel composite insulation for Mars pressures (4–8 torr) k is either 0.012 or 0.005 W/m K with thicknesses of about 0.40 and 0.32 m, respectively. For an average temperature difference of 200 K on Mars, the heat flow per unit area for is $k \times 200$/(thickness). For foam and aerogel composites, the resultant heat leaks are 6.0 W/m^2 for foam and 3.1 W/m^2 for aerogel composites. These heat leak rates will lead to boil-off rate of 61 and 32% per month, respectively. Furthermore, it is not clear how feasible it would be to use aerogel composites on very large tanks.

A.2.4 Effect of Boil-Off in a Closed System

This section is based on the work of Mueller et al. (1994). The concept behind the liquid storage option is as follows: during coast periods, a parasitic heat leak into a liquid hydrogen tank causes liquid boil-off and consequently a tank pressure rise. (It is assumed that uniform liquid and vapor temperatures are maintained by a mixer). The temperature of the tank contents (liquid and vapor) also rises to maintain saturation conditions. Models indicate that the energy absorption capability of the liquid is dominant, so the net effect is that the pressure rises more slowly for high liquid fill-levels. Pressure-time traces for different initial liquid fraction fill levels are shown in Fig. A.3 for a tank size (10 m^3) with a heat leak (10 W) starting at 1 atm pressure (101.3 MPa or 14.7 psia). Note that Mueller et al. used a tank with surface area 22.4 m^2, so that their heat leak was 0.45 W/m^2. This is toward the low end of other estimates. The results show that as energy is added to the tank, the liquid fraction either rises so that eventually the tank fills with liquid, or it falls so that the tank fills with vapor (if no venting were to occur) depending on the initial fraction of the tank filled with liquid. It is important to note that while the rates of pressure rise for high liquid fractions are smaller, once the tank fills with liquid, the pressure rises extremely rapidly and the tank will fail. Another important observation is that for a given maximum tank design pressure, venting will be required more often towards the end of a mission when the liquid level is lower rather than at the beginning of the mission when the liquid level is high. These results apply only if the contents of the tank are well mixed. If there is significant thermal stratification, the tank pressure in general will rise more quickly, and the higher liquid levels will not necessarily have slower pressure rises than the lower liquid levels. The experience gained in designing a mixer for the Space Shuttle Centaur upper stage has shown that very compact and lightweight mixers can be developed for hydrogen-fueled upper stages. The pressure-control benefits of such mixers far outweigh their mass and power penalties such that a mixer will almost certainly be included on any hydrogen-fueled upper stage.

A.2.5 Zero Boil-Off Concepts

An alternate approach to using a larger tank and allowing boil-off is to use a tank that contains the ultimate amount of hydrogen needed from the start, and provide an active cooling system to remove an equivalent amount of heat that continually leaks in during the storage period, thus avoiding boil-off. This is called the "zero boil-off" (ZBO) approach. The ZBO approach eliminates the need for a larger tank, but it requires a cryo-cooler, a controller, heat transfer equipment in the tank, and a power system to provide power. The greater complexity of this system implies more risk than for passive storage systems. Furthermore, ZBO technology is relatively immature and those who advocate it tend to be optimistic regarding heat leak rates. Plachta and Kittel (2003), Kittel and Plachta (1999) and Kittel et al. (1998) dealt with ZBO techniques for transporting cryogens to Mars and long-term storage in LEO. Panzarella et al. (2003) modeled ZBO systems. The trade involved in deciding whether ZBO is appropriate depends on the comparison of the larger tank loaded with hydrogen at the start for the passive storage case versus the smaller tank with ZBO plus the cryo-cooler, radiator, controller and power system required to implement ZBO. Plachta and Kittel (2003) performed such a comparison. For the ZBO case, they estimated the masses of the tank, insulation, cryocooler, solar array and the radiator. They found that for sufficiently long storage durations, ZBO has lower mass. For hydrogen, the cross-over duration in LEO where the estimated masses for ZBO and non-ZBO masses are equal varies from about 90 days for a 2-m diameter tank to about 60 days for a 3.5-m diameter tank to about 50 days for a 5-m diameter tank. For the hydrogen tank described previously, with a diameter of 3.3 m and a fully loaded mass of ~ 1525 kg, they estimate that the mass of the additional equipment to implement ZBO

Fig. A.3 Pressure rise versus
liquid volume fraction as a
function of initial liquid
volume fraction and time. For
initial liquid volume fractions
of 0.5 or above, the tank
eventually becomes filled
with liquid, whereas for initial
liquid volume fractions less
than 0.5, the tank eventually
fills with vapor

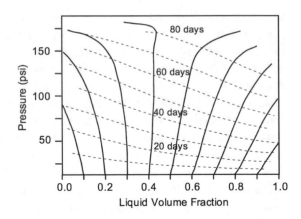

is roughly 175 kg so the ZBO storage system has a mass of ~ 1700 kg for all durations. The required mass of the non-ZBO storage system starts out at 1525 kg for zero duration and increases to 1700 kg for a duration of 62 days, and 2210 kg for 270 days duration. Thus for long-term storage, the ZBO approach appears to have significant advantages, based on the estimated heat leak of 0.5 W/m^2. Plachta and Kittel (2003) would therefore predict that hydrogen represents $1250/1700 = 73.5\%$ of the total mass using ZBO. The surface area of this tank is 34.2 m^2, so the estimated heat leak is likely to be somewhere around 17 W based on 0.5 W/m^2. According to Guernsey et al. (2005), it would typically require 500 W of electrical power to remove 1 W of heat from a hydrogen tank at 20 K. Using ZBO, the electrical power requirement would appear to be 8.5 kW. Plachta and Kittel (2003) estimated that this could be provided within a mass of 175 kg. This depends on how the power is generated. If a nuclear reactor is required, the mass might be considerably higher. Furthermore, if we used a higher heat leak than 0.5 W/m^2, the power requirement would rise and the mass estimate of 175 kg for the ZBO system would be an underestimate. Haberbusch et al. (2004) were concerned with long-term storage of cryogens for periods of 1–10 years. They analyzed a spherical liquid hydrogen storage tank, with a passive multilayer insulation system, and an actively cooled shield system using a cryo-cooler. The study investigated the effects of tank size, fluid storage temperature (densification), number of actively cooled shields, and the insulation thickness on the total system mass, input power, and volume. The hydrogen temperature was 21 K and the ullage allowance was 2%. They considered a tank that could be cooled by two cryo-coolers, with the inner shield cooled to either 16 or 21 K, and the outer shield cooled to 80 K with a uniform external temperature of 294 K. The analysis was carried out for three cases: (1) 250 or 4000 kg of liquid hydrogen stored without an outer shield (2) 250 kg of liquid hydrogen storage with an 80 K outer shield, and (3) 4000 kg of liquid hydrogen storage with an 80 K outer shield. They assumed that the effective conductivity model can be extrapolated to 300 layers since it is fairly linear. Experimentally derived factors to model additional heat leak from MLI degradation due to seams and penetrations and for conduction through supports and pipes resulted from testing. The MLI factor as defined in the model was the ratio of heat leak due to seams and penetrations to the heat transfer rate through the blanket. The conduction factor was the ratio of non-insulation heat transfer to heat transfer rates through the insulation system. The MLI factor was 1.74 and the conduction factor was 0.14. The MLI factor is a strong function of thickness and seam type, but not temperature. Therefore the assumption was made that the same MLI factor can be used regardless of the boundary temperatures used in the analyses for MLI with up to 300 layers. They found significant benefits from cooling the inner shield, and even greater benefits from cooling the inner and outer shields. The inner shield could be cooled to either 16 K or 21 K, and the outer shield could be cooled to 80 K using active cryo-coolers. In each model, the power and mass requirements were estimated for a ZBO system to match the heat loads, although they never specified what the heat loads were. Haberbusch et al. (2004) found that when only the inner shield was cooled, the power and mass requirements for zero boil-off were very high with a moderate thickness of MLI but decreased significantly with increasing blanket thickness. For a tank holding 4000 kg of hydrogen (3 m diameter by 9 m length) use of blankets with 100 layers of MLI resulted in a total storage system mass of 2000 kg, a power requirement of 10 kW, and a volume requirement of about 60 m^3 (only slightly greater than the actual hydrogen volume of about 57 m^3). The surface area of the tank was about 86 m^2. Using the rule-of-thumb from Guernsey et al. that it takes 500 W of power to remove 1 W of heat from a liquid hydrogen tank, the apparent heat leak into the tank is estimated at 10,000/500 = 20 W, or 0.23 W/m^2. The model of Guernsey et al. (2005) (including 80% margin and allowance for penetrations and conduction) would predict that for 100 layers of MLI, the heat leak would be about 0.36 W/m^2. Using cooling of only the inner shield, hydrogen represents 4000/6000 = 67% of the total mass. In the case where both the inner shield and the outer shield were actively cooled, for the 4000 kg hydrogen tank, the storage system mass minimizes for about 75 layers of MLI in the outer blanket, and decreases as the inner blanket is made thicker. For an inner blanket of ∼150 layers and an outer blanket of ∼75 layers, the estimated storage total mass is 1750 kg, the power level is ∼2 kW, and the volume is ∼64 m^3. Hydrogen would then represent about 4000/5750 = 70% of the total mass. These are very attractive figures if they can be achieved.

A.3 Storage as a Dense Gas at Reduced Temperature

The requirements for a cryo-cooler to remove heat leaks at 80–120 K are far less demanding than for 20–30 K. Therefore it is worth considering storage of hydrogen in this temperature range. Figure A.4 shows the variation of density of gaseous hydrogen with pressure at various temperatures. Supercritical hydrogen can have about the same density as liquid hydrogen under various conditions of pressure and temperature. For example, hydrogen gas at 90 K and 5000 psi, or 120 K and 7000 psi, has the same density as liquid hydrogen at 100 psi. Storing hydrogen as a gas at such super-critical pressures has some attractive features because liquid-vapor separation is not needed and the hydrogen can be pressure-fed to the user with no need for pumps or compressors. However, the mass of

Fig. A.4 Variation of density of gaseous hydrogen with pressure at various temperatures

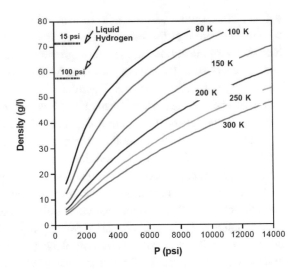

such a tank would be determined mainly by the operating pressure, and as we have seen previously, the hydrogen content of a storage system at these pressures is expected to be 5–7%. This would be unattractive for transfers to the Moon or Mars.

A.4 Storage as Solid Hydrogen

Solid hydrogen storage presents several complications: (1) difficulties in supplying hydrogen to users at a high enough pressure and flow rate. (2) mass penalties associated with a full-volume internal metal foam, and (3) additional complexity of the ground loading system. Delivery of hydrogen to users when stored as a sublimating solid will be difficult. A source of heat can be used to sublime some of the solid hydrogen but hydrogen vapor must be constantly removed to keep the tank below the triple point pressure (7.0 kPa = 0.07 bar = 1.0 psi) and prevent the solid from melting. A compressor (basically a vacuum pump) is needed that has an inlet (suction) pressure lower than the 1.0 psi triple point pressure. Whether such a compressor is used to supply end-users directly or to charge accumulators, there are significant issues associated with the flow rate and/or the pressure ratio required, as well issues of leakage past pistons, seals, etc. Power requirements could be significant. It may be possible to use metal hydrides to pressurize the hydrogen. Requirements do not seem to have been studied. Regardless of the approach selected to add energy to sublime hydrogen, some method is needed to hold the solid hydrogen in place. An aluminum foam with a relative density of 2% (2% of the density of a solid chunk of aluminum) has a density of about 56 kg/m^3. Since the density of solid hydrogen is 86.6 kg/m^3, the foam add significantly to the mass of the tank. The complexity of the ground support system and the cost of operations for solidifying hydrogen in large storage tanks appears to be formidable. The method used to fill the solid hydrogen is likely to involve circulation of liquid helium around the liquid-hydrogen-filled inner vessel to freeze the hydrogen. This is a very expensive method. The fact that the pressure within the hydrogen tank is only ~1 psi introduces concerns about structural stability of the tank and air ingestion into the tank.

A.5 Storage as Solid-Liquid Slush

The use of slush hydrogen (SLH) at the triple point pressure (7.042 kPa = 1.02 psia) provides several advantages and several disadvantages compared to liquid hydrogen (Friedlander et al. 1991), On the positive side, a 50% solid mass fraction in slush has 15% greater density and 18% greater heat capacity than normal boiling point liquid hydrogen. The increase in density allows a more compact tank, and the increased heat capacity reduces the amount of venting required for a given heat leak. Slush also has the advantage that the solid mass fraction can be allowed to vary to accommodate heat leaks without venting. During storage, heat leaks cause some of the solid in the slush to melt, reducing the solid mass fraction (some of the vapor condenses as well to maintain the tank at the triple point pressure). However, it is not clear how to perform vapor extraction with vapor at 1 psi. A thermodynamic vent system, with its Joule-Thomson expansion device, would become clogged by the solid hydrogen particles in the slush. While the vapor extraction problem may possibly be surmountable, the low vapor pressure of slush hydrogen (the triple point pressure) presents the same problems in compressor design as for the solid storage concept. In addition, the ground handling of slush is complicated by the fact that the volume of the slush increases as heat enters and the solid melts. As heat leaks into the slush, some of the solid will melt and the overall volume of the slush will increase. If left unchecked, the volume of the slush will exceed the volume of the tank and it will come out the vent. If the slush melted completely, it would occupy the volume of the same mass of liquid. Making the tank large enough to accommodate this (or only filling it partially) negates the density/volume savings of using slush in the first place. Thus slush conditioning will be required on the pad to maintain adequate solid fractions as heat leaks into the tank. Many of the ground handling concerns have been addressed in the National Aerospace Plane (NASP) design effort but developing the infrastructure for slush handling at launch facilities will be very expensive. According to Hardy et al. (1992):

> The production of densified hydrogen does not appear to be a difficult task. Several production methods for SLH have been demonstrated at the laboratory level, and densifying hydrogen to a triple-point liquid is easily achieved by pumping to reduce the fluid to triple-point pressure and temperature. The SLH production methods include evaporative cooling processes such as 'freeze-thaw', … and refrigeration processes …. The major issues related to production of SLH, include large-scale production, methods of production, safety concerns, and energy efficiency and costs….

A.6 Storage as Hydrogen at Its Triple Point

Triple-point hydrogen (TPH)—liquid hydrogen at 1.02 psia and 13.8 K, offers increases in density (8%) and heat capacity (12%) in comparison to liquid H_2 at its normal boiling point. Although these benefits are not as large as those for slush, TPH does not have the added complication of solid particles, and thus may also be an option for future space vehicles.

A.7 Storage as Adsorbed Hydrogen on a Sorbent

Gas/sorbent pairs are characterized by whether sorption takes place by physical adherence due to polarization forces ("physi-sorption"), or whether the gas molecules chemically combine with the sorbent ("chemi-sorption"). The great advantage of chemi-sorption is that stronger gas-surface forces are involved whereas physi-sorption involves weaker van der Walls forces. With these stronger forces, greater storage densities of gases can be achieved. However, it is difficult to find a suitable chemi-sorption sorbent material for a gas like hydrogen. Dissociative hydride alloys have been used but they tend to be too massive and expensive for use in commercial storage systems. The advantage of physi-sorption sorbent materials is that they can be used for many gases, particularly hydrogen. All these choices are challenged by the demands of high gravimetric and volumetric density for any conceivable application. Finely powdered ("activated") carbon is a widely used physi-sorption material used for absorbing and desorbing gases. It derives its effectiveness from the very high surface area that can be produced if processed effectively. Typically, the adsorption energy (ε) to a sorbent is a few 10 s of meV, and because of the weak forces, physi-sorption of hydrogen works best at cryogenic temperatures. Since adsorption scales with ε/kT, retaining good adsorption of hydrogen on carbon up to room temperature requires the adsorption energy ε to be increased to about 200 meV. Therefore, the "holy grail" in physi-sorption systems for effective storage of hydrogen at room temperatures is to find ways to structure the sorbent surfaces to increase the effective adsorption energy ε to about 200 meV.

In the decade since their discovery, carbon nano-tubes have become a focal point of the nano-materials field, and many hopes have been pinned on them. Scientists have envisaged these molecular-scale graphitic tubes as the key to a variety of potentially revolutionary technologies ranging from super-strong composites to nano-electronics. Some conjecture that carbon nano-tubes might provide a good medium for storing hydrogen. When carbon nano-tubes first became available, they drew interest because these tubes are typically produced in bundles that are lightweight and have a high density of small, uniform, cylindrical pores (the individual nano-tubes). Under the right conditions, there is no reason why nano-tubes would not allow hydrogen molecules into their interior space or into the channels between the tubes. But the crucial question is this: Can nano-tubes store and release practical amounts of hydrogen under reasonable conditions of temperature and pressure? An early report by chemist Nelly M. Rodriguez and co-workers at Northeastern University reported that certain graphite nano-fibers can store hydrogen at levels exceeding 50 wt% at room temperature (Park et al. 1998). The density was claimed to be higher than that of liquid hydrogen. Such results are incredible when one considers that methane has a hydrogen content of only 25 wt% and it is hard to imagine packing more hydrogen around carbon atoms than in methane. Attempts at other labs to reproduce the Northeastern findings were unsuccessful. A year later, Jianyi Lin and coworkers in the physics department at the National University of Singapore reported remarkable hydrogen uptake by alkali-metal-doped multi-walled nano-tubes formed by the catalytic decomposition of methane. The H_2 uptake was claimed to be 20 wt% for lithium-doped nano-tubes at 380 °C and 14 wt% for potassium-doped nanotubes at room temperature. Subsequent studies in other labs cast doubt on these results, attributing them to the presence of water impurities. Both the Northeastern and Singapore work are now regarded as aberrations due poor experimental technique. The era of exaggerated claims for nano-carbon storage of hydrogen is now hopefully past, but the question remains as to how much hydrogen such materials can possibly hold. The answer to that question depends on whom you ask. The most optimistic view is that such materials may be able to store up to 8% by weight of hydrogen at room temperature and moderate pressure. Others claim that the hydrogen uptake could be as high as 6 wt% in nano-tubes—but at only 77 K, the temperature of liquid nitrogen. At room temperature, it is claimed that nano-tubes adsorb only minor amounts of hydrogen.

A.8 Storage on Metal Hydrides

This section is based partly on Colozza (2002). Metal hydrides are metallic alloys that absorb hydrogen. These alloys can be used as a storage medium because of their ability to not only absorb hydrogen but also to release it. The release of hydrogen is directly related to the temperature of the hydride. Typically metal hydrides can hold hydrogen equal to approximately 1–2% of their weight. If the temperature is held constant the hydrogen is released at a constant pressure. The metal hydride tank can be used repeatedly to store and release hydrogen cyclically. The limiting factor on its ability to store hydrogen is the accumulation of impurities within the tank. These impurities fill the spaces that would normally store the hydrogen and thereby reduce the tanks capacity. While metal hydrides are very useful in some terrestrial applications, they do not hold enough hydrogen by weight percentage to be viable for space transfer of hydrogen.

Appendix B
Recent Advances in RWGS Technology

This brief Appendix was inserted in the proof pages as a late entry based on an email received from Robert Zubrin on November 30, 2017. Further experimental work at Pioneer Energy was reported on the RWGS reaction:

$$CO_2 + H_2 ==> CO + H_2O$$

A 99+% conversion efficiency was achieved at a scale that produced about 3 g/h of H_2O, which could then be used to produce almost 3 g/h of oxygen by electrolysis. That would amount to about 75 g/sol, and 30 kg of oxygen over 400 sols. This is an important development and it is hoped that further details will eventually be released.

In addition, Pioneer Energy fed the CO produced by the RWGS reaction into a methanol synthesis unit where liquid fuel was produced.

© Springer International Publishing AG 2018
D. Rapp, *Use of Extraterrestrial Resources for Human Space Missions to Moon or Mars*, Springer Praxis Books, https://doi.org/10.1007/978-3-319-72694-6

Appendix C
New NASA Program for ISRU
Development—December 4, 2017

In December 2017, NASA announced a bold, new program to further develop ISRU technology.[2] This solicitation for proposals should bring forth exciting new work in ISRU in the period 2019 to 2022. This NASA announcement occurred while my book was in press, and I inserted this Appendix as a brief summary and commentary on the NASA program. I will refer to the NASA Broad Agency Announcement as "the BAA".

As we discussed earlier in this book, the NASA plan for a human mission to Mars involves a crew of four, and the current NASA assessment of the need for ascent propellants is "approximately 30 metric tons of oxygen and methane".

The BAA will consider proposals in the following categories:

Trade Studies: NASA is interested in funding a limited number of trade studies to help identify critical architecture and technology gaps and to further define the benefits of ISRU for various destinations and mission profiles. The emphasis should be on comparing and contrasting different options (e.g., a mission with and without ISRU, comparing one ISRU option to another ISRU option for a given mission, etc.). What seems to be missing here is studying the requirements and approaches for prospecting for resources. This seems to be a pervasive omission in almost all the NASA Mars studies I have seen. While the BAA (and more generally NASA) want to develop technology for utilizing various kinds of soils ranging from granular to consolidated and icy soils, and extracting water from such sources, there doesn't seem to be much interest in space missions to locate good regions for exploitation, and validation of extraction processes in situ. The methods for doing this might involve significant new technology innovation. As I said in the Preface to this book:

> This might involve balloons, airplanes or gliders, network landers, or possibly an orbiter that dips down to low altitudes for brief periods. None of these technologies seem to be high on NASA's priority list.

Component Development: This includes development and testing of components unique to ISRU in a relevant environment, where such hardware is not available in state-of-the-art hardware.

---

[2]Next Space Technologies for Exploration Partnerships-2 (NextSTEP-2), Appendix D: In-Situ Resource Utilization (ISRU) Technology Broad Agency Announcement NNH16ZCQ001K-ISRU December 4, 2017.

© Springer International Publishing AG 2018
D. Rapp, *Use of Extraterrestrial Resources for Human Space Missions to Moon or Mars*, Springer Praxis Books, https://doi.org/10.1007/978-3-319-72694-6

Component and Subsystem Development: This part of the program is aimed at development and testing of ISRU subsystems that include critical component(s). Subsystems of interest include atmosphere collection, oxygen or fuel production, excavation, and soil processing. Testing in a relevant environment is required.

One of the difficulties in preparing a BAA of this sort is understanding at the outset what technology advances might be possible, and how far they might be developed within a given time period and budget. NASA has a tendency to "push the envelope" by asking for perhaps more than seems likely to be delivered, hoping that the doing organization might contribute some cost sharing. In defining the state of maturity of development, NASA relies on the "Technology Readiness Level" (TRL) scale. The BAA focuses on TRL 5 and TRL 6 with some work at the TRL 4 level. Unfortunately, the meaning of the terse language in this scale[3] is subject to a wide variety of interpretations. And in many cases, the steps of the TRL scale do not fit the actual maturation process for technology. Of particular importance in the present connection is the definition of TRL 5 and TRL 6. While the official definitions are:

TRL 5 = Component and/or breadboard validation in a relevant environment.
TRL 6 = System and/or subsystem model or prototype demonstration in a relevant environment (ground or space).

While there is fairly wide acceptance on the difference between a component, a subsystem and a system, it is not at all clear what the differences are between a breadboard, a model and a prototype. One possible definition (of many) is:

- A breadboard has almost all the performance attributes of a real system, but it is not packaged into a compact, rugged, robust unit. Instead, it might spread out on a board or boards with gangling wires and tubes.
- A model (usually called "engineering model") or a prototype has essentially all the form, fit and function of a flight model but does utilize flight-qualified parts.

Whether NASA would agree with these definitions is anyone's guess.

Within the BAA's component development track, the plan is to make several awards in the range $250K to $500K per year in three one-year phases, ending with a TRL 5 "hardware demonstration" after three years. Based on experience gained from working on MOXIE from 2014 through 2017, it is clear that any such testing system will require a significant chamber, instrumentation, and data logging. The amounts of funding appear to be inadequate, although a group that has been actively working on a component for some time might have a good deal of the equipment in place prior to BAA funding. Nevertheless, running such experiments requires considerable human effort as well.

Within the BAA's component/subsystem development track, the plan is to make several awards in the range $250K to $750K per year in two phases, achieving TRL 5 in

---

[3]https://www.nasa.gov/sites/default/files/trl.png

18 months, and ending with a TRL 6 "subsystem demonstration" two years later. The allocation of funding and duration seem very skimpy to this writer.

One other aspect of the BAA is that the technical approach must be described in 10–15 pages within a proposal, which can only be described as ridiculous. Is it possible that NASA already knows who and what it wants to fund, and is issuing this BAA pro forma as a means of providing that funding? If such were the case, a 10-page proposal would be more than adequate.

Despite the misgivings expressed in previous paragraphs, I have great hope that this program will advance the state of ISRU in significant ways.

References

Abbud-Madrid, Angel, et al. 2016. Mars Water In-Situ Resource Utilization (ISRU) Planning (M-WIP) Study. Report of the Mars Water In-Situ Resource Utilization (ISRU) Planning (M-WIP) Study; 90 p, posted April, 2016 at http://mepag.nasa.gov/reports/Mars_Water_ISRU_Study.pptx.

Adamo, Daniel R., and James S. Logan. 2016. Aquarius, a reusable water-based interplanetary human spaceflight transport. *Acta Astronautica* 128: 160–179.

Adan-Plaza, S., et al. 1998. Extraction of Atmospheric Water on Mars. http://www.lpi.usra.edu/publications/reports/CB-955/washington.pdf.

Adler, Mark, et al. 2009. Entry, Descent and Landing Roadmap. https://www.nasa.gov/pdf/501326main_TA09-EDL-DRAFT-Nov2010-A.pdf.

Allen, C.C., et al. 1998. Martian Regolith Simulant JSC Mars-1. Lunar and Planetary Science XXIX, Paper 1690.

Anonymous. 2005. Exploration Systems Architecture Study (ESAS), NASA-TM-2005-214062. http://www.sti.nasa.gov, November 2005.

Anonymous. 2006. Project Constellation Presentation on Mars Mission Architectures, attributed to D. Cooke, NASA deputy associate administrator for exploration.

Ash, R.L., W.L. Dowler, and G. Varsi. 1978. Feasibility of Rocket Propellant Production on Mars. *Acta Astronautica* 5: 705–724.

Augustynowicz, S.D., J.E. Fesmire, and J.P. Wikstrom. 1999. Cryogenic insulation systems. 20th International Refrigeration Congress, Sydney. In *Cryogenic Insulation System for Soft Vacuums*, ed. S.D. Augustynowicz, and J.E. Fesmire. Montreal CEC.

Barlow, N.G., and T.L. Bradley. 1990. Martian Impact Craters: Dependence of Ejecta and Interior Morphologies on Diameter, Latitude, and Terrain. *Icarus* 87: 156–179.

Bidrawn, F. 2008. The Effect of Ca, Sr, and Ba Doping on the Ionic Conductivity and Cathode Performance of LaFeO$_3$. *Journal of the Electrochemical Society* 155: B660–B665.

Blair, Brad R., et al. 2002. Space Resource Economic Analysis Toolkit: The Case for Commercial Lunar Ice Mining. Final Report to the NASA Exploration Team, December 20, 2002.

Bonin, Grant, and Tarik Kaya. 2007. End-to-End Analysis of Solar-Electric-Propulsion Earth Orbit Raising for Interplanetary Missions. *Journal of Spacecraft and Rockets* 44: 1081–1093.

Bradshaw, Jeffery M. 2008. *Preliminary Report of the Small Pressurized Rover*. http://www.jeffreymbradshaw.net/publications/SPR-ReportDraft.pdf.

Braun, R.D., and R.M. Manning. 2006. Mars Exploration Entry, Descent and Landing Challenges. In 2006 IEEE Aerospace Conference, IEEEAC paper #0076. Montana: Big Sky.

Brooks, K.P., S.D. Rassat and W.E. TeGrotenhuis. 2005. Development of a Microchannel ISPP System PNNL Report, PNNL-15456, September, 2005.

© Springer International Publishing AG 2018
D. Rapp, *Use of Extraterrestrial Resources for Human Space Missions to Moon or Mars*, Springer Praxis Books, https://doi.org/10.1007/978-3-319-72694-6

Bruckner, A., et al. 1998. Extraction of Atmospheric Water on Mars for the Mars Reference Mission. HEDS-UP Mars Exploration Forum, Lunar and Planetary Institute, Houston, TX, May 4–5, 1998.

Calle, C.I. 2011. Dust Removal Technology for a Mars In Situ Resource Utilization System. In *AIAA 2011-7348*.

Caltech. 2017. Lunar Extraction for Extra-terrestrial Prospecting. Caltech Space Challenge.

Carpenter, Christian, et al. 2012. A Sustainable Logistics Architecture for Exploration of the Moon, Asteroids, and Mars. http://www.rocket.com/files/aerojet/documents/Architecture/GLEX_2012055_9x12266.pdf.

Cassady, R. Joseph, et al. 2015. Next Steps in the Evolvable Path to Mars. 66th International Astronautical Congress, Jerusalem, Israel, IAC-15-D2,8-A5.4,8.

Chamberlain, M.A., and W.V. Boynton. 2004. Modeling Depth to Ground Ice on Mars. *Lunar and Planetary Science* XXXV.

Christian, John A., Grant Wells, Jarret Lafleur, Kavya Manyapu, Amanda Verges, Charity Lewis and Robert D. Braun. 2006. Sizing of an Entry, Descent, and Landing System for Human Mars Exploration. In *Georgia Institute of Technology, AIAA 2006-7427, AIAA Space 2006 Conference*, September 2006, San Jose, CA.

Clancey, William J. 2004. Participant Observation of a Mars Surface Habitat Mission Simulation. http://cogprints.org/3967/1/Habitation04Clancey.pdf.

Clark, D.L. 1997. In-Situ Propellant Production On Mars: A Sabatier/Electrolysis Demonstration Plant. In *AIAA 97-2764* (July).

Clark, D.L. 1998. Progress Reports to JPL on Sorption Compressor; D. Rapp, JPL Contract Manager.

Clark, D.L. 2003. Test Results. LMA memo to J. Badger, JSC, October 8, 2003.

Clark, D.L., and K. Payne. 2000. Evaluation of Mars CO_2 Acquisition Using CO_2 Solidification. Report to NASA on Contracts IWTA# 7IDO362ETAC9HECEP1A5 and IWTA# 7IDO362E-TAC0HECEP1A5, May 30, 2000.

Clark, D.L., et al. 1996. In Situ Propellant Production Demonstration. Final Report MCR-95-561, JPL Contract 960247, Lockheed Martin Astronautics, Boulder, CO, March 22, 1996.

Clark, D.L., K.S. Payne and J.R. Trevathan. 2001. Carbon Dioxide Collection and Purification System for Mars. In *AIAA 2001-4660*.

Cohen, M.M. 2000. Pressurized Rover Airlocks. SAE Technical Series 2000-01-2389.

Collins, Daniel L. 1985. Psychological Issues Relevant to Astronaut Selection for Long-Duration Space Flight: A Review of the Literature. http://www.dtic.mil/dtic/tr/fulltext/u2/a154051.pdf.

Colozza, Anthony J. 2002. Hydrogen Storage for Aircraft Applications Overview. NASA/CR—2002-211867, September, 2002.

Coons, S., R. Curtis, C. McLain, J. Williams, R. Warwick and A. Bruckner. 1995. ISPP Strategies and Applications for a Low Cost Mars Sample Return Mission. In *AIAA 95-2796*.

Coons, S., J. Williams and A. Bruckner. 1997. Feasibility Study of Water Vapor Adsorption on Mars for In-Situ Resource Utilization. In *AIAA 97-2765*.

Crow, S.C. 1997. The MOXCE Project: New Cells for Producing Oxygen on Mars. In *AIAA 97-2766*, July 1997.

Crow, S.C., and K. Ramohalli. 1996. Theoretical analysis of Zirconia Cell Performance. Personal communication, April 28, 1996.

Curtis, Howard. 2005. *Orbital Mechanics for Engineering Students*. Elsevier Butterworth–Heinemann Linacre House, Jordan Hill, Oxford OX2 8DP.

Cutright, Bruce L. and William Ambrose. 2014. Economic Benefits of In Situ Resource Utilization in Near-Term Space Exploration. www.searchanddiscovery.com/documents/2014/70163cutright/ndx_cutright.pdf.

Deirmendjian, D. 1964. Scattering and polarization properties of water clouds and hazes in the visible and infrared. *Applied Optics* 3: 187–196. (1969). *Electromagnetic Scattering on Spherical Polydispersions*, 75–119. New York: Elsevier.

DRM-1, Hoffman, Stephen J. and David I. Kaplan (eds.). 1997. Human Exploration of Mars: The Reference Mission of the NASA Mars Exploration Study Team. Lyndon B. Johnson Space Center, Houston, Texas, July 1997, NASA Special Publication 6107.

DRM-3, Drake, B.G. (eds.). 1998. Reference Mission Version 3.0 Addendum to the Human Exploration of Mars: The Reference Mission of the NASA Mars Exploration Study Team. NASA/SP-06107-ADD, Lyndon B. Johnson Space Center. June 1998.

DRM-5, Drake, B.G. (ed.). 2009. Design Reference Architecture Version 5.0, Lyndon B. Johnson Space Center, July 2009.

Dudley-Rowley, Marilyn, et al. 2002. Crew Size, Composition, and Time. In *AIAA-2002-6111*.

Ebbesen, Sune Dalgaard, and Mogens Mogensen. 2009. Electrolysis of Carbon Dioxide in Solid Oxide Electrolysis Cells. *Journal of Power Sources* 193: 349–358.

Eckart, Peter. 1996. *Parametric Model of a Lunar Base for Mass and Cost Estimates*. Wissenschaft: Herbert Utz Verlag.

Finn, J.E., C.P. McKay, and K.R. Sridhar. 1996. Martian Atmospheric Utilization by Temperature-Swing Adsorption. In *26th SAE International Conference on Environmental Systems*. Monterey, CA, July 8–11, 1996.

Friedlander, Alan, Robert Zubrin, and Terry L. Hardy. 1991. Benefits of Slush Hydrogen for Space Missions. NASA TM 104503, October 1991.

Genta, Giancarlo. 2015. Human Mars Missions: Why? How? When? Recent Advances in Fluid Mechanics and Thermal Engineering. http://www.wseas.us/e-library/conferences/2015/Salerno/MECH/MECH-03.pdf.

Graps, Amara, et al. 2017. In Space Utilization of Asteroids" ASIME 2016: Asteroid Intersections with Mine Engineering, Luxembourg. September 21–22, 2016.

Gueret, C., M. Daroux, and F. Billaud. 1997. Methane Pyrolysis: Thermodynamics. *Chemical Engineering Science* 52: 815–827.

Guernsey, Carl S., Raymond S. Baker, David Plachta, Co-Investigator, Peter Kittel, Robert J. Christie, and John Jurns. 2005. Cryogenic Propulsion With Zero Boil-Off Storage Applied To Outer Planetary Exploration, Final Report, April 8, 2005, (JPL D-31783).

Haberbusch, Mark S., Robert J. Stochl, and Adam J. Culler. 2004. Thermally Optimized Zero Boil-off Densified Cryogen Storage System for Space. *Cryogenics* 44: 485–491.

Haberle, R.M., et al. 1993. Atmospheric Effects on the Utility of Electric Power on Mars. *Resources of Near-Earth Space*. London: University of Arizona Press.

Hardy, Terry L. and Margaret V. Whalen. 1992. Technology Issues Associated With Using Densified Hydrogen for Space Vehicles. NASA TM 105642. In *AIAA 92-3079, AIAA/SAE/ASME/ASEE 28th Joint Propulsion Conference*. Nashville, TN, July 1992.

Heavens, N.G., et al. 2011. The Vertical Distribution of Dust in the Martian Atmosphere During Northern Spring and Summer. *Journal of Geophysical Research* 116: E04003.

Hecht, M.H. et al. 2017. MOXIE, ISRU, and the History of In Situ Studies of the Hazards of Dust in Human Exploration of MARS. https://www.hou.usra.edu/meetings/marsdust2017/pdf/6036.pdf.

Hirata, Christopher, et al. 1999. A New Plan for Sending Humans to Mars: The Mars Society Mission. http://www.lpi.usra.edu/publications/reports/CB-979/caltech99.pdf.

Hoffman, Stephen, et al. 2016. "Mining" Water Ice on Mars: An Assessment of ISRU Options in Support of Future Human Missions. https://www.nasa.gov/sites/default/files/atoms/files/mars_ice_drilling_assessment_v6_for_public_release.pdf.

Hofstetter, Wilifried K., Paul D. Wooster and Edward F. Crawley. 2007. Analysis of Human Lunar Outpost Strategies and Architectures. In *AIAA 2007-6726*.

Hofstetter, Wilfried K. 2006. Extending NASA's Exploration Systems Architecture Towards Long term Crewed Moon and Mars Operations. In *AIAA 2006-5746*.

Holladay, J.D., K.P. Brooks, R. Wegeng, J. Hua, J. Sanders, and S. Baird. 2007. Microreactor Development for Martian In situ Propellant Production. *Catalysis Today* 120: 35–44.

Hornbeck, G. et al. 2003. HUMEX A Study on the Survivability and Adaptation of Humans to Long-Duration Exploratory Missions. ESA Report SP-1264.

Jakosky, Bruce M., Michael T. Mellon, E. Stacy Varnes, William C. Feldman, William V. Boynton, and Robert M. Haberle. 2005. Mars Low-Latitude Neutron Distribution: Possible Remnant Near-Surface Water Ice and a Mechanism for Its Recent Emplacement. *Icarus* 175: 58–67; Erratum: *Icarus* 178: 291–293.

Kanas, Nick. 2014. Psychosocial Issues During an Expedition to Mars. *Acta Astronautica* 103: 73–80.

Kikuchi, E., and Y. Chen. 1997. Low Temperature Syngas Formation by CO_2 Reforming of Methane in a Hydrogen-Permselective Membrane Reactor. Natural Gas Conversion IV. *Studies in Surface Science and Catalysis* 107: 547–553.

Kim-Lohsoontorn, Pattaraporn, and Joongmyeon Bae. 2011. Electrochemical performance of solid oxide electrolysis cell electrodes under high-temperature coelectrolysis of steam and carbon dioxide. *Journal of Power Sources* 196: 7161–7168.

Kittel, P., and D. Plachta. 1999. Propellant Preservation for Mars Missions," Advances in Cryogenic Engineering, Vol. 45, presented at the Cryogenic Engineering Conference, Montreal, Canada, July 12–16, 1999. In *Cryocoolers for Human and Robotic Missions to Mars*, ed. Kittel, P., L. Salerno, D. Plachta, Cryocoolers 10, presented at the 10th International Cryocooler Conference, Monterey, California, May 1998.

Kleinhenz, Julie E. and Aaron Paz. 2017. An ISRU Propellant Production System to Fully Fueled a Mars Ascent Vehicle. In *AIAA 2017-0423*.

Lamamy, Julien-Alexandre, et al. 2005. Balancing Humans and Automation for the Surface Exploration of Mars. IAC-05-A5.2.02.

Landau, Damon F., and James M. Longuski. 2009. Comparative assessment of human–Mars-mission technologies and architectures. *Acta Astronautica* 65: 893–911.

Landis, G.A. 1996. Dust obscuration of mars solar arrays. *Acta Astronautica* 38: 885–891.

Larson, William E. 2011. NASA's ISRU Project Status for FY2012 and Beyond. http://ntrs.larc.nasa.gov/search.jsp?N=4294932186+4294504246&Nn=4294967061|Document+Type|Technical+Report||4294965276|Subject+Terms|CLOSED+ECOLOGICAL+SYSTEMS.

Larson, William E., and Gerald B. Sanders. 2010. The In-Situ Resource Utilization Project: under the New Exploration Enterprise. http://ntrs.larc.nasa.gov/search.jsp?N=4294932186+4294504246&Nn=4294967061|Document+Type|Technical+Report||4294965276|Subject+Terms|CLOSED+ECOLOGICAL+SYSTEMS.

Larson, William E., Gerald B. Sanders and Kurt R. Sacksteder. 2010. NASA's In-Situ Resource Utilization Project: Current Accomplishments and Exciting Future Plans. http://ntrs.larc.nasa.gov/search.jsp?N=4294932186+4294504246&Nn=4294967061|Document+Type|Technical+Report||4294965276|Subject+Terms|CLOSED+ECOLOGICAL+SYSTEMS.

Lin, Jianyi, et al. 1999. Remarkable hydrogen uptake by alkali-metal. *Science* 285: 91.

Linne, D.L., et al. 2013. Demonstration of Critical Systems for Propellant Production on Mars for Science and Exploration Missions. In *AIAA 2013-0587*.

Linne, D.L., G. Sanders and K. Taminger. 2015. Capability and Technology Performance Goals for the Next Step in Affordable Human Exploration of Space. In *AIAA 2015-1650*.

Linne, D. 2016. Overview of HEOMD Advanced Exploration Systems ISRU Technology Development Project. Presented at the MOXIE Science Team Meeting November 2–4, 2016.

Makri, M., Y. Jiang, I.V. Yentekakis, and C.G. Vayenas. 1996. Oxidative coupling of methane to ethylene with 85% yield in a gas recycle electrocatalytic or catalytic reactor-separator. In *11th International Congress on Catalysis—40th Anniversary, Studies in Surface Science and Catalysis*, vol. 101, ed. A. Part, 387–396. Amsterdam: Elsevier.

Manzey, D. 2004. Human missions to Mars: New psychological challenges and research issues. *Acta Astronautica* 55: 781–790.

Mark, M.F., and W.F. Maier. 1996. CO_2 reforming of methane on supported Rh and Ir catalysts. *Journal of Catalysis* 164: 122–130.

Marsh, Christopher L., and Robert D. Braun. 2009. Fully-Propulsive Mars Atmospheric Transit Strategies for High-Mass Payload Missions. IEEEAC paper #1219.

McKay, Mary, David McKay and Michael Duke. 1992. Space Resources – Materials, NASA Report SP-509, vol. 3.

McLean, John, et al. 2017. Testing The Mars 2020 Oxygen In-Situ Resource Utilization Experiment (Moxie) HEPA Filter and Scroll Pump in Simulated Mars Conditions. *Lunar and Planetary Science* XLVIII: #2410.

Mellon, M.T., and B.M. Jakosky. 1995. The distribution and behavior of Martian ground ice during past and present epochs. *Journal of Geophysical Research* 100: 11781–11799.

Merrell, R.C. 2007. Microchannel ISPP as an Enabling Technology for Mars Architecture Concepts. In *AIAA 2007-6055*.

Meyer, P.J. 1996. A Theoretical Analysis of the Performance of Solid Electrolyte Cells for Oxygen Production from Carbon Dioxide. M. S. Thesis, Professor K. Ramohalli, University of Arizona.

Morphew, M.E. 2001. Psychological and human factors in long duration spaceflight. *McGill Journal of Medicine* 6: 74–80.

Moss, Shaun. 2013. Blue dragon! An affordable and achievable Humans-to-Mars mission architecture and program to create an International Mars Research Station. https://shaunmoss.com/Blue%20Dragon%20-%20MASTER%20v2.pdf.

Mueller, Paul J., Brian G. Williams and J.C. Batty. 1994. Long-term hydrogen storage and delivery for low-thrust space propulsion systems. In *AIAA-1994-3025, 30th ASME, SAE, and ASEE, Joint Propulsion Conference and Exhibit*. Indianapolis, IN, June 27–29, 1994.

Muscatello, A.C., et al. 2012. Mars In Situ Resource Utilization Technology Evaluation. In *AIAA 2012-0360*. 2014. Atmospheric Processing Module for Mars Propellant Production. *Earth and Space* 2014: 444–454.

Musk Elon. 2017. Making Humans a Multi-Planetary Species" *New Space*, June 2017, **5**, 46-61. https://doi.org/10.1089/space.2017.29009.emu.

NASA. 2011. NASA's Analog Missions: Paving the way for Space Exploration. NP-2011-06-395-LaRC.

NASA. 2015. NASA's Journey to Mars: Pioneering Next Steps in Space Exploration. https://www.nasa.gov/sites/default/files/atoms/files/journey-to-mars-next-steps-20151008_508.pdf.

Nørnberg, P., H.P. Gunnlaugsson, J.P. Merrison, and A.L. Vendelboe. 2009. Salten Skov I: A Martian magnetic dust analogue. *Planetary and Space Science* 57: 628–631.

Noyes, G.P. 1988. Carbon Dioxide Reduction Processes for Spacecraft ECLSS: A Comprehensive Review. SAE Technical Paper 881042, 18th Intersociety Conference on Environmental Systems, San Francisco, July 11–13, 1988.

Noyes, G.P., and R.J. Cusick. 1986. An Advanced Carbon reactor for Carbon Dioxide Reduction. SAE Technical Paper 860995, July, 1986.

Palac, Don, et al. 2015. Nuclear Systems Kilopower Project Overview. Presented at Nuclear and Emerging Technologies for Space (NETS) 2015 Conference, Albuquerque, New Mexico, February, 2015, http://anstd.ans.org/wp-content/uploads/2015/07/5024_Palac-et-al.pdf.

Palinkas, L.A. 2001. Psychosocial issues in long-term space flight: overview. *Gravitational and Space Biology* 14: 25–33.

Panzarella, Charles H., and Mohammad Kassemi. 2003. Simulations of Zero Boil-Off in a Cryogenic Storage Tank. In *41st Aerospace Sciences Meeting and Exhibit*. January 6–9, 2003, Reno, Nevada (*AIAA 2003-1159*).

Park, A.C., T.K. Baker, and N.M. Rodriguez. 1998. Hydrogen Storage in Graphite Nanofibers. *Journal of Physical Chemistry B* 102: 4253–4256, May 28, 1998.

Perez-Davis, Marla E. and K.A. Faymon. 1987. Mars Manned Transportation Vehicle. NASA Technical Memorandum 101487, July 1987.

Petrova, E.V., et al. 2011. Optical Depth of the Martian Atmosphere and Surface Albedo From High Resolution Orbiter Images. http://www-mars.lmd.jussieu.fr/paris2011/abstracts/petrova_paris2011.pdf.

Plachta, David and Peter Kittel. 2003. An Updated Zero Boil-Off Cryogenic Propellant Storage Analysis Applied to Upper Stages or Depots in an LEO Environment. NASA/TM—2003-211691, June 2003. In *AIAA–2002–3589*.

Pollack, J.B., R.M. Haberle, J. Schaffer, and H. Lee. 1990. Simulations of the General Circulation of the Martian Atmosphere I. Polar Processes. *Journal of Geophysical Research* 95: 1447–1473.

Polsgrove, T., et al. 2015. Mars Ascent Vehicle for Human Exploration. https://ntrs.nasa.gov/search.jsp?R=20160006401.

Polsgrove, T., et al. 2017. Human Mars Ascent Vehicle Configuration and Performance Sensitivities. https://www.researchgate.net/publication/317702039_Human_Mars_ascent_vehicle_configuration_and_performance_sensitivities.

Ponelis, A.A., and P.G.S. van Zyl. 1997. CO_2 Reforming in Methane with a Membrane Reactor. Natural Gas Conversion IV. *Studies in Surface Science and Catalysis* 107: 555–560.

Portree, David S.F. 2001. Humans to Mars: Fifty Years of Mission Planning, 1950—2000, NASA History Division, Office of Policy and Plans, NASA Headquarters, Washington, DC 20546. *Monographs in Aerospace History, Series*, Number 21, February 2001.

Ramohalli, K.N.R. 1991. Technologies of ISRU/ISMU. 42nd Congress of the International Astronautical Federation, IAA-91-659, Montreal, Canada, October 1991.

Rapp, D. 2006. Mars life support systems. *Mars Journal*. http://marsjournal.org/contents/2006/0005/files/rapp_mars_2006_0005.pdf.

Rapp, D. 2015. *Human Missions to Mars*, 2nd ed. Heidelburg: Springer/Praxis.

Rapp, D., P. Karlmann, D.L. Clark and C.M. Carr. 1997. Adsorption Pump for Acquisition and Compression of Atmospheric CO_2 on Mars. In *AIAA 97-2763*, July, 1997.

Rapp, D., G. Voecks, P. Sharma and N. Rohatgi. 1998. Methane Reforming for Mars Applications. JPL Report D-15560, March 30, 1998; also presented at 33rd IECEC, 1998.

Rhatigen, Jennifer L. 2011. *Constellation Program Lessons Learned, Volume I: Executive Summary*. NASA Report SP-2011-6127.

Richter, R. 1981. *Basic Investigation Into the Production of Oxygen in a Solid Electrolyte*. In *AIAA-81-1175, AIAA 16th Thermophysics Conference*, June 23–25, 1981, Palo Alto, CA.

Rowland, Chad, et al. 2004. *Surface Mobility Technology Development: Pressurized Mars Rovers*. https://www.semanticscholar.org/paper/Surface-Mobility-Technology-Development-Pressurize-Rowland-Paulson/9817b1603cad9efd98b8737f766451d4860ef6be.

Rucker, Michelle. 2013. *Mars Surface Drilling Study, DRA 5.0*. JSC Report 66635.

Rucker, M.A. 2016. *Surface Power for Mars*. Presented at FISO, 12-7–16, http://spirit.as.utexas.edu/%7Efiso/telecon/Rucker_12-7-16/.

Rucker, M.A., et al. 2015. *Integrated Surface Power Strategy for Mars*. Presented at Nuclear and Emerging Technologies for Space (NETS) 2015 Conference, Albuquerque, New Mexico, February, 2015, https://ntrs.nasa.gov/search.jsp?R=20150000526.

Rucker, M.A., et al. 2016. *AIAA-2016-5452*. Solar vs. Fission Power for Mars. Presented at AIAA SPACE 2016, Long Beach, September, 2016, https://ntrs.nasa.gov/search.jsp?R=20160002628.

Sacksteder, Kurt R. and Gerald B. Sanders. 2007. In-Situ Resource Utilization for Lunar and Mars Exploration. In *AIAA 2007-345*.

Sadler Machine Co. 2014. An Overview of Recent and Future Lunar/Mars Habitat Terrestrial Analogs. http://www.agrospaceconference.com/wp-content/uploads/2014/06/Pres_ASC_2014_Sadler_s5.pdf.

Salotti, Jean-Marc, et al. 2012. http://www.planete-mars.com/wp-content/uploads/2012/12/Simple-roadmap.pdf.

Salotti, Jean-Marc, et al. 2014. Crew Size Impact on the Design, Risks and Cost of a Human Mission to Mars. IEEE Aerospace Conference Proceedings.

Sandal, Gro Mjeldheim. 2001. Psychosocial issues in space: Future challenges. *Gravitational and Space Biology* 14: 47–54.

Sanders, Gerald B. 2005a. A Case for In-Situ Resource Utilization (ISRU) for Human Mars Exploration. Presentation 11-18-2004.

Sanders, Gerald B. 2005b. In-Situ Resource Utilization (ISRU) Capabilities & Roadmapping Activities. LEAG/SRR Meeting South Shore Harbor, League City, TX, Oct 26, 2005.

Sanders, Gerald B. 2005c. Project Plan for In-Situ Resource Utilization (ISRU).

Sanders, G.B. 2012. Mars ISRU—Update from 2004 Mars Human Precursor SSG Study, presentation. http://ntrs.larc.nasa.gov/search.jsp?N=4294932186+4294504246&Nn=4294967061|Document+Type|Technical+Report||4294965276|Subject+Terms|CLOSED+ECOLOGICAL+SYSTEMS.

Sanders, G.B. 2016. In Situ Resource Utilization (ISRU) Theme Group Presentation for Human Exploration of Mars. Presented at the MOXIE Science Team Meeting November 2–4, 2016.

Sanders, J., and M. Duke. 2005. ISRU Capability Roadmap Team Final Report. NASA Informal report edited by J. Sanders and M. Duke, March 2005.

Sanders, Gerald B., Thomas Simon, William E. Larson, Edgardo Santiago-Maldonado, Kurt Sacksteder, Diane Linne, John Caruso, and Robert Easter. 2007. ISRU at a Lunar Outpost: Implementation and Opportunities for Partnerships and Commercial Development. http://ntrs.larc.nasa.gov/search.jsp?N=4294932186+4294504246&Nn=4294967061|Document+Type|Technical+Report||4294965276|Subject+Terms|CLOSED+ECOLOGICAL+SYSTEMS.

Sanders, G.B. and W.E. Larson. 2011. Progress Made in Lunar In-Situ Resource Utilization under NASA's Exploration Technology and Development Program. NASA Technical Report.

Sanders, G.B., W.E. Larson and M. Picard. 2011a. Development and Demonstration of Sustainable Surface Infrastructure for Moon/Mars Missions. 62nd International Astronautical Congress 2011 Cape Town, South Africa Oct. 5, 2011.

Sanders, G.B., et al. 2011b. RESOLVE for Lunar Polar Ice/Volatile Characterization Mission. EPSC Abstracts vol. 6, EPSC-DPS2011-1605.

Schlacht, Irene Lia, et al. 2016. Space Analog Survey: Review of Existing and New Proposal of Space Habitats with Earth Applications. 46th International Conference on Environmental Systems, ICES-2016-367.

Schwarz, M. 1996. *Thermal Analysis of a Solid Oxide Electrolysis System for Oxygen Production on Mars*. M.S. Thesis, Professor K. Ramohalli, University of Arizona.

Sedlak, J.M., J.F. Austin, and A.B. LaConti. 1981. Hydrogen recovery and purification using the solid polymer electrolyte electrolysis cell. *International Journal of Hydrogen* 6: 45–51.

Sercel, Joel. 1986. Solar Thermal Propulsion for Planetary Spacecraft. https://ntrs.nasa.gov/search.jsp?R=19860007953; https://ntrs.nasa.gov/search.jsp?R=19860000381.

Sercel, Joel. 2015. APIS (Asteroid Provided In-Situ Supplies): 100MT Of Water from a Single
    Falcon 9. https://www.nasa.gov/feature/apis-asteroid-provided-in-situ-supplies-100mt-of-water-
    from-a-single-falcon-9.
Sercel, Joel. 2016. Asteroid Provided In-situ Supplies (APIS): A Breakthrough to Enable an
    Affordable NASA Program of Human Exploration and Commercial Space Industrialization.
    NIAC Phase I Final Report, 29 February 2016.
Sercel, Joel. 2017. Optical Mining of Asteroids, Moons, and Planets to Enable Sustainable Human
    Exploration and Space Industrialization. https://www.nasa.gov/directorates/spacetech/niac/2017_
    Phase_I_Phase_II/Sustainable_Human_Exploration.
Sridhar, K.R. 1995. Personal Communication to D. Rapp, November 6, 1995.
Sridhar, K.R., and B.T. Vaniman. 1995. Oxygen Production on Mars Using Solid Oxide
    Electrolysis. 25th International Conference on Environmental Systems, Paper 951737, San
    Diego, July 10–13, 1995.
Sridhar, K.R., and S.A. Miller. 1994. Solid oxide electrolysis technology for ISRU and life support.
    Space Technology 14: 339–346.
Stancati, M.L., J.C. Niehoff, W.C. Wells and R.L. Ash. 1979. Remote Automated Propellant
    Production: A New Potential for Round Trip Spacecraft. In AIAA 79-0906, May 1979.
Steinfeldt, B.L., et al. 2009. High Mass Mars Entry, Descent, and Landing Architecture Assessment.
    In AIAA-2009-6684.
Stewart, Sarah T., Thomas J. Ahrens, and John D. O'Keefe. 2004. Impact-Induced Melting of
    Near-Surface Water Ice on Mars. In 13th APS Topical Conference on Shock-Compression of
    Condensed Matter—2003, ed. M.D. Furnish, Y.M. Gupta, and J.W. Forbes, 1484–1487. New
    York: The American Institute of Physics.
Stuster, Jack W. 2000. Bold endeavors: Behavioral lessons from polar and space exploration.
    Gravitational and Space Biology Bulletin 13: 48–57.
Suitor, J.W., D.J. Clark and Losey R.W. 1990. Development of Alternative Oxygen Production
    Source Using a Zirconia Solid Electrolyte Membrane. JPL Publication D-7790, Jet Propulsion
    Laboratory, Pasadena, CA, August, 1990.
Szynkiewicz, A., et al. 2010. Origin of terrestrial gypsum dunes—Implications for Martian
    gypsum-rich dunes of Olympia Undae. Geomorphology 121: 69–83.
Tao, G., K.R. Sridhar, and C.L. Chan. 2004. Electrolysis of carbon dioxide in solid oxide
    electrolysis cells. Solid State Ionics 175: 621–624.
Taylor, Lawrence A., and Thomas T. Meek. 2005. Microwave sintering of lunar soil: Properties,
    theory, and practice. Journal of Aerospace Engineering 18: 188–196.
Tomasko, M.G., et al. 1999. Properties of Dust in the Martian Atmosphere from the Imager on Mars
    Pathfinder. Journal of Geophysical Research—Planet 104: 8987–9007.
Vakoch, Douglas A. (ed.). 2011. Psychology of Space Exploration: Contemporary research in
    historical perspective, NASA History Series, NASA Report SP-2011-4411.
van Keulen, A.N.J., M.E.S. Hegarty, J.R.H. Ross and P.F. Van den Oosterkamp. 1997. The
    Development of Platinum-YSZ Catalysts for the CO_2 Reforming of Methane. Natural Gas
    Conversion IV, Studies on Surface Science and Catalysis 107: 537–546.
Wells, G., J. Lafleur, A. Verges, K. Manyapu, J. Christian, C. Lewis and R. Braun. 2006. Entry,
    Descent and Landing Challenges of Human Mars Exploration. 29th AAS Guidance and Control
    Conference February 2006, Breckenridge, CO, AAS 06-072.
Wickman, Leslie A. 2006. Human Performance Considerations for a Mars Mission. IEEE Aerospace
    Conference Proceedings.
Williams, J., S. Coons, and A. Bruckner. 1995. Design of a water vapor adsorption reactor for
    martian in-situ resource utilization. Journal of the British Interplanetary Society 48: 347–354.

Wilson, R.J., and M.I. Richardson. 2000. The Martian Atmosphere During the Viking Mission. *Icarus* 145: 555–579.

Wood, S.E., and D.A. Paige. 1992. Modeling the Martian Seasonal CO_2 Cycle 1. Fitting the Viking Lander Pressure Curves. *Icarus* 99: 1–14.

Woodcock, Gordon. 2004. Controllability Of Large SEP For Earth Orbit Raising. In *40th AIAA/ASME/ SAE/ASEE Joint Propulsion Conference and Exhibit*. 11–14 July 2004, Fort Lauderdale, Florida, AIAA 2004-3643.

Zacny, Kris. 2017. Asteroid ISRU. http://www.lpi.usra.edu/sbag/meetings/jan2017/presentations/Zacny.pdf.

Zubrin, R. 1999. *Entering Space*. New York: Jeremy P. Tarcher/Putnam.

Zubrin, R.M., and D.B. Weaver. 1993. Practical Methods for Near-Term Piloted Mars Mission. American Institute of Aeronautics and Astronautics. In *AIAA-93-20898*.

Zubrin, R.M., D.A. Baker and O. Gwynne. 1991. Mars Direct: A Simple, Robust, and Cost Effective Architecture for the Space Exploration Initiative. American Institute of Aeronautics and Astronautics. In *AIAA-91-0328*.

Zubrin, R., S. Price, L. Mason and L. Clark. 1994. Report on the Construction and Operation of a Mars In-Situ Propellant Production Plant. In *AIAA-94-2844, 30th AIAA/ASME Joint Propulsion Conference*. Indianapolis, IN, June, 1994.

Zubrin, R., S. Price, L. Mason and L. Clark. 1995. Mars Sample Return with In-Situ Resource Utilization: An End to End Demonstration of a Full Scale Mars In-Situ Propellant Production Unit. NASA Contract No. NAS 9-19145. Presented to NASA January 13, 1995.

Zubrin, R., M.B. Clapp and T. Meyer. 1997a. New Approaches for Mars ISRU Based on the Reverse Water Gas Shift. In *AIAA-97-0895*, January, 1997.

Zubrin, R., B. Frankie and T. Kito. 1997b. Mars In-Situ Resource Utilization Based on the Reverse Water Gas Shift. In *AIAA-97-2767, 33rd AIAA/ASME Joint Propulsion Conference*, Seattle, WA, July 6–9, 1997.

Zubrin, R., B. Frankie and T. Kito. 1998. Report on the Construction and Operation of a Mars In Situ Propellant Production Unit Utilizing the Reverse Water Gas Shift. In *AIAA-98-3303, 34th AIAA/ASEE Joint Propulsion Conference*, July 13–15, 1998, Cleveland Ohio.

Index

© Springer International Publishing AG 2018
D. Rapp, *Use of Extraterrestrial Resources for Human Space Missions to Moon or Mars*, Springer Praxis Books, https://doi.org/10.1007/978-3-319-72694-6

Printed in the United States
By Bookmasters